DATE DUE

5-30-97

Antibody Engineering

....

Chemical Immunology

Vol. 65

Series Editors
Luciano Adorini, Milan
Ken-ichi Arai, Tokyo
Claudia Berek, Berlin
J. Donald Capra, Dallas, Tex.
Anne-Marie Schmitt-Verhulst, Marseille
Byron H. Waksman, New York, N.Y.

Basel · Freiburg · Paris · London · New York ·
New Delhi · Bangkok · Singapore · Tokyo · Sydney

Antibody Engineering

Volume Editor   *J. Donald Capra,* Dallas, Tex.

43 figures, 1 in color, and 9 tables, 1997

KARGER   Basel · Freiburg · Paris · London · New York ·
New Delhi · Bangkok · Singapore · Tokyo · Sydney

······················

Chemical Immunology

Formerly published as 'Progress in Allergy'
Founded 1939 by Paul Kallòs

Library of Congress Cataloging-in-Publication Data
Antibody engineering / volume editor, J. Donald Capra.
p. cm. — (Chemical immunology; 65)
Includes bibliographical references and index.
1. Immunoglobulins – Biotechnology.   2. Protein engineering.
I. Capra, J. Donald, 1937–.   II. Series.
TP248.65.I49A572   1997
615′.37–dc20
ISBN 3–8055–6356–6 (hbk.)

Bibliographic Indices. This publication is listed in bibliographic services, including Current Contents® and Index Medicus.

Contents

Preface XI

**Engineering Minibody-Like Ligands by Design and
Selection**
M. Sollazzo, S. Venturini, S. Lorenzetti, M. Pinola, F. Martin, Pomezia   1

The Why and How of Designing Proteins   1
Design of the Minibody Framework   2
Phage Display: In vitro Molecular Evolution   6
Minibody Repertoire   7
Affinity-Selection of a Minibody Antagonist of Human IL-6   7
Selection of Minibodies with Serine Protease Inhibitor Activity   10
Outlook   12
Acknowledgments   14
References   15

**Antibodies against HIV-1 from Phage Display Libraries:
Mapping of an Immune Response and Progress towards
Antiviral Immunotherapy**
P.W.H.I. Parren, D.R. Burton, La Jolla, Calif.   18

Antibody Phage Display Libraries   20
Expression of Whole Antibody Molecules   23
Antibodies to Viruses   24
Neutralization of Virus by Recombinant Fab Fragment and Whole IgG   25
Antibodies to HIV-1   27

Antibodies to the CD4-Binding Site of gp120 29
  Neutralization of Unpassaged Plasmavirus 32
  Protection against HIV-1 Infection in hu-PBL-SCID Mice by Induction of
    Passive Immunity 33
Antibodies to Other Epitopes on gp120 35
  Epitope Masking 35
  Panning on Peptides 38
Antibodies to gp41 39
In vitro Evolution of Fabs to Improve Affinity and Strain Cross-Reactivity and
    Increase Neutralization Potency 41
Recognition Properties of Fab b12 Indicate Efficient Binding to Native HIV-1
    Envelope Oligomer 43
Antibody Phage Display Libraries as Tools to Assess Vaccines 45
Conclusions 47
Acknowledgments 48
References 48

Chemical Engineering at the Antibody Hinge

G.T. Stevenson, Southampton 57

Essential Anatomy of the Antibody Molecule 57
The Immunoglobulin Hinge 58
Functional Types of SS Bonds 59
Manipulating Cysteine Residues 60
  SS Interchange 61
  Alkylation by Maleimides 62
  Oxidation of SH Groups 63
The Fab'γ Module 64
  F(ab')$_2$ 64
  Fab'(SH)$_5$ 64
  Fab'-SS-Py 64
  Fab'-SH and Fab'-Maleimide 66
The Fcγ Module 66
  Fcγ1 from Basic IgG 66
  Fc-SS-Py 67
  Fc-SH and Fc-Maleimide 68
Antibody Constructs 68
  Planning the Constructs 68
  A Sample Construct: Bispecific Fab$_2$Fc$_2$ (Anti-CD20/Anti-CD37) 69
References 71

Structure and Function in IgA

J.M. Hexham, L. Carayannopoulos, J.D. Capra, Dallas, Tex.   73

Biochemistry of IgA   74
    Structure of IgA   74
    Polymerization of IgA   76
    Secretory IgA   77
    Effector Functions of IgA   79
    IgA Comparative Biology   80
Molecular Analysis of IgA   81
Future Directions   82
References   83

IgG Effector Mechanisms

M.R. Clark, Cambridge   88

IgG Isotypes   89
IgG Functions   93
Recombinant IgG Antibodies   95
Residues Affecting Complement Binding   97
Residues Affecting High-Affinity FcγRI (CD64) Binding   99
Residues Affecting Low-Affinity FcγRII (CD32) Binding 100
Residues Affecting ADCC through FcγRIII (CD16) 101
Glycosylation and Effector Functions 102
Neonatal Transport and the Role of FcRn 103
Is IgG Catabolism Related to Binding to FcRn? 104
Acknowledgments 106
References 106

Glycosylation of Antibody Molecules:
Structural and Functional Significance

R. Jefferis, J. Lund, Birmingham 111

Antibody Glycosylation 112
Structural Consequences of IgG Glycosylation 114
Functional Consequences of Asn297 Glycosylation 115
Factors Influencing Glycosylation Hybridoma, and Recombinant Immunoglobulin
    Molecules 121
References 125

Engineering Novel Antibody Molecules

M.G. Sensel, M.J. Coloma, E.T. Harvill, S.-U. Shin, R.I.F. Smith,

S.L. Morrison, Los Angeles, Calif. 129

Antibody Fusion Proteins 131
   Antibody Fusion Proteins for Receptor-Mediated Targeting of the Brain 133
     Binding of Fusion Proteins to Their GF Receptors 135
     Effector Functions of the IgG3-C_H3-GF Fusion Proteins 136
     In vivo Targeting of the GF Fusion Proteins to the Brain 137
   IgG3-IL-2 Fusion Proteins 139
     The IgG3-IL-2 Fusion Protein Stimulates Both Proliferation and
     Cytotoxicity 141
     The IgG3-IL-2 Fusion Protein Binds to High- and Intermediate-Affinity
     IL-2 Receptors 141
     IgG3-IL-2 Binds FcγRI and Activates C′ 142
     In vivo Properties of IgG3-IL-2 142
     Effectiveness of the IgG3-IL-2 Fusion Protein in Eliciting a Tumor-Specific
     Immune Response 143
     Potentiation of Host Antibody Response by αDNS-IgG3-IL-2-Bound
     Antigen 145
Polymeric IgG Molecules 146
   Production of IgM-Like Mutants of IgG 146
   Analysis of Complement Activation by Polymeric IgG 148
   FcγR Binding by Polymeric IgG 149
   In vivo Half-Life of Polymeric IgG 149
Conclusions and Future Directions 151
Acknowledgments 154
References 154

Ligand Function of Antigenized Antibodies Expressing the RGD Motif

R. Billetta, P. Lanza, M. Zanetti, La Jolla, Calif.

. . . . 159

The Process of Antigenization 161
Materials and Methods 161
   Computer-Based Analysis and Selection of the Engineered Sequence 161
   Antigenization of Antibodies with the RGD Motif 163
   Immunochemical Analysis of AgAbs 165
   Adhesion and Inhibition Assay 165
Studies on the Antigenicity of Antibodies Engineered to Express Arg-Gly-Asp . . . 166
Discussion 172
   Considerations for Peptide Structure 172
   Considerations for Tumor and Angiogenesis 173
Acknowledgments 175
References 176

Immunogenicity of Viral Epitopes Expressed on Genetically and Enzymatically Engineered Immunoglobulins

C.A. Bona, A. Bot, T.-D. Brumeanu, New York, N.Y. 179

Immunoglobulins as Delivery Systems 180
   Antibody-Peptide Constructs Prepared by Chemical Methods 180
   Antibody-Peptide Constructs Prepared by Enzymatically-Glycosidic-
     Mediated Conjugation 182
   Chimeric Immunoglobulins Expressing Foreign Peptides or Biologically
     Active Ligands 184
Chimeric Immunoglobulins Expressing Peptides Recognized by T Cells 188
   Immunogenicity of Peptides Expressed in Chimeric Immunoglobulins 190
   Immunopotency of HA110–120 Peptide Expressed in Chimeric Molecules . . . 193
   Kinetics of Generation and Persistence of Viral Peptide Expressed in
     Various Carrier Molecules 193
   Molecular Mechanisms Responsible for the Presentation of Viral Epitopes
     Expressed in Chimeric Immunoglobulin Molecules 197
Conclusions 201
References 203

Subject Index 207

Preface

The unequivocal linking of immunity with substances contained in the blood emerged from the findings in the latter part of the 19th century which showed that the sera from animals immunized with bacterial toxins protected other animals, on passive transfer, from the otherwise lethal effects of the toxin. For almost all of the 20th century, antibody research has relied on immunoglobulins that were produced in animals. These molecules were used to either study the structure of antibodies, or for therapeutic purposes in elegant extensions of the original work described above. In the 1950s, 'gammaglobulin' became a mainstay in the treatment of immunodeficiency diseases, and in the last decade both intravenous immunoglobulin and mouse and human hybridomas have entered the arena of human diagnosis and therapeutics. During the last decade of the 20th century, a remarkable revolution is taking place; antibodies can now be directly engineered to produce both specificities and effector functions that are either novel or better than starting materials. This remarkable advance, while still in its infancy, has begun to bear considerable fruit in both investigative and therapeutic modalities. The discovery of recombinant DNA was quickly applied to antibody molecules and now, antibodies can be engineered with various structures so that detailed information to correlate structure and function is a matter of course. Thus, rather than the laborious process of isolating fragments and subfragments of immunoglobulins from various species, one can now produce highly purified domains or subdomains to study function in a direct and unequivocable fashion.

Similarly, the variable regions of antibody molecules have been engineered such that structures can be swapped between molecules from different species and mutations can be introduced to both increase affinity, broaden or narrow specificity. In the broadest of contexts, the term 'antibody engineering' is used whenever one uses molecular biological procedures to modify or alter antibody molecules.

This volume of *Chemical Immunology* grew out of my sense that it was timely to bring together, in one volume, the varying approaches to antibody engineering that have been extant for a number of years. As the newly appointed editor in chief of the series, it seemed appropriate to me that the first volume that I would specifically edit would cover this field.

The volume is roughly divided into three parts. The first part includes papers from the laboratories of Drs. Sollazzo and Burton and involves engineering antibody variable regions. Over the last few years, Dr. Sollazzo and his colleagues have elegantly shown the extraordinary properties of the 'minibody' and how it can be used to both search for ligands as well as provide critical structural information about the molecules themselves. This is followed by a chapter from Dennis Burton's laboratory in which they describe the use of phage display libraries to map immune responses against a critical human pathogen, the human immunodeficiency virus.

Moving toward the constant region, the chapter from Dr. Stevenson's laboratory describes newer approaches to chemical engineering of the antibody hinge. Relatively few studies have been done on this obviously crucial part of antibody molecules. The hinge is responsible for both the flexibility of the antibody molecule and is thought by some to transmit the signal indicating that antigen has been engaged in the combining site. This signal is thought to initiate some of the biological properties of the Fc region.

Three studies on constant regions follow. One from my own laboratory on engineering IgA molecules, one from Roy Jefferis' laboratory describing the glycosylation of antibody molecules and one from Sherie Morrison's laboratory detailing engineered novel antibody molecules and the role of various structures in the Fc as they pertain to clearance. These three chapters elegantly show the application of antibody engineering to the Fc region of antibodies and how by a clever use of molecular biology, one can generate structures that have novel properties and can be used to deduce basic fundamental aspects of structure-function analysis.

Finally, chapters from Drs. Zanetti and Bona describe the use of antibody molecules to deliver antigens to the immune system in a unique and powerful way by genetically engineering various motifs within the variable regions of immunoglobulin molecules. Powerful immune responses of both the humoral and cellular nature can thus be enlisted. These chapters review this area and provide clear approaches for the future.

I am grateful to my colleagues for delivering these manuscripts in a timely manner and for their quick responses to my editorial queries throughout the publication process. I'm also grateful to the staff at Karger for providing outstanding assistance in the publication of this book.

J. Donald Capra

Capra JD (ed): Antibody Engineering.
Chem Immunol. Basel, Karger, 1997, vol 65, pp 1–17

....

Engineering Minibody-Like Ligands by Design and Selection

Maurizio Sollazzo, Sara Venturini, Stefano Lorenzetti, Mari Pinola, Franck Martin

Istituto di Ricerche di Biologia Molecolare (IRBM) P. Angeletti, Pomezia, Italy

The Why and How of Designing Proteins

The ability to engineer polypeptides and proteins that can carry out specific functions is radically transforming the way we think about basic science, pharmacology and biotechnology. One can easily predict that designed proteins will have a profound impact upon the conception of novel drugs, the utilization of natural resources and engineering biopolymers with unique properties not found in nature. Proteins with tailored functions can be designed either by modifying existing molecules or, in principle, by inventing entirely new structures and sequences that do not exist in nature. While considerable progress in de novo design has been made, a protein has yet to be designed that possesses all of the characteristics of a natural protein [1]. The challenge for protein designers arises partly from the unmanageable number of possible sequences that can be conceived and partly from the unimaginably small fraction of these that can be studied experimentally or computationally. Fortunately, not all the possible sequence combinations need to be considered because many different sequences can adopt the same structure. Studying protein sequences that are simpler allow us to focus attention on the principal determinants of structure and can teach us some of the fundamental principles governing protein folding. Recent examples of this strategy are the simplification of protein surface and cores [for a review, see ref. 2], and the use of a binary 'code' to design α-helical proteins [3]. The design of partially degenerate sequences combined with biological selection and amplification results in an even more powerful tool. Given our present limited understanding of the rules that regulate protein folding, the most feasible route for developing proteins with desirable biological functions combines rational design with experimental screening and genetic selection schemes [4].

Many groups, including our own, have been involved in establishing strategies that allow the 'in vitro evolution' of polypeptide sequences as a strategy for creating polypeptides with predetermined activity [5, and references therein]. Our approach to developing β-proteins with novel activities is that of combining the rational design of a convenient framework (minibody) with extensive mutagenesis of targeted regions to generate large repertoires of β-pleated molecules with variable surfaces composed of β-turns. The choice of using an antibody domain as a model stems from these molecules representing the most diverse natural molecular recognition system known and from the profound understanding of structure-function relationship accumulated over the past decade for these domains. Antibody variable domains are unrivalled instruments for generating molecular diversity. A simplified representation of these domains is six partially randomized constrained loops mounted onto a rigid β-barrel framework. By finding the right equilibrium between sequence randomization of the loops and constraint to allow for 'induced fit' plus alternative packing geometry of their variable domains, antibodies optimize structural diversity and generate virtually all possible surfaces for binding any antigen [6]. We designed the 'minibody' with the objective of reproducing at least some of the extraordinary structural diversity of antibodies in a smaller polypeptide framework that would enable us to produce specific ligands for a given target. We wanted to take advantage of some of the features – typical of antibody variable domains – such as tolerance to sequence variability and predictability of the main chain conformation of the hypervariable regions (canonical structures) [7]. The reasons for such an undertaking will become clearer in the following section.

Design of the Minibody Framework

Five years ago, we conceived the idea of constructing a 'minimalist' version of an antibody variable domain that had only two hypervariable loops (fig. 1), yet was still capable of producing specific ligands with adequate affinity. All attractive interactions between macromolecules are dominated by electrostatic and hydrophobic forces; in addition, hydrogen bonding and shape complementarity play a crucial role. The highest possible affinity towards a cognate molecule would be achieved by a ligand displaying a perfect mirror image of the shape of the target surface, with a charge distribution perfectly complementary to that of the target molecules. A major factor limiting the affinity and specificity of interactions is conformational flexibility. Polypeptide ligands

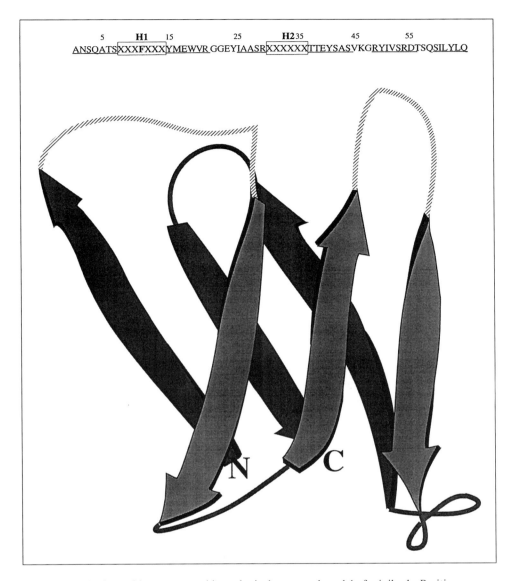

Fig. 1. Amino acid sequence and hypothetical structural model of minibody. Positions corresponding to the H1 and H2 hypervariable loops are shown within boxes, β-strands are underlined. In the model, β-strands are depicted as arrows, whereas the two hypervariable regions are represented as dashed loops.

that can easily undergo induced fit will be able to interact with a variety of targets, but will pay the price of decreased affinity due to a conformational energy penalty. On the contrary, molecules with a rigid conformation that naturally complement the target surface will bind to fewer targets, but will render higher affinities. In other words, the more a peptide segment is constrained, the less likely it is to bind to any particular target, but at the same time, if it does bind, binding is likely to be tighter and more specific. We felt that if we could design the minibody polypeptide this would be a satisfactory compromise between two extremes. The particular advantages over full antibodies that our format would offer is that of having two constrained loops embedded in a small protein scaffold, allowing us to focus interaction between ligand and ligate to relatively few residues. Because antibodies have six complementarity-determining regions (CDRs), it is virtually impossible to direct antigen binding to only one or two CDRs with perhaps a few exceptions (see last section).

We engineered the minibody by focusing on the re-design of mouse antibody variable domain V_H McPC603. That this was a feasible objective had been suggested by an earlier report showing that one subdomain portion of V_H McPC603 was sufficient to assemble effectively with its cognate light (L) chain and bind to the phosphorylcholine hapten [8]. The outcome of the design was a 61-residue protein consisting of three β-strands from each of the two β-sheets of the heavy (H) chain variable domain along with the H1 and H2 hypervariable loops [9, 10]. In order to facilitate the study of the biochemical properties of our first-generation molecule produced by solid-phase synthesis [11], we constructed a second generation of molecules (by recombinant DNA techniques) suitable for expression in the insoluble fraction of *Escherichia coli* cytoplasm [12]. This bacterial expression system facilitated further engineering of the minibody and permitted the metabolic-labeling of the molecule for isotope-edited NMR spectroscopy. The first problem we tackled was to improve the water solubility of the parental molecule, MB1, which was 14 μM, the same magnitude as that of V_H domains: such a poor water solubility is probably a consequence of hydrophobic amino acid side chains being exposed to the solvent. We subsequently slightly modified the minibody framework and obtained variants with higher solubility in aqueous buffers but with lesser or equal thermodynamic stability than MB1 [12].

The results of far-UV circular dichroism (CD) analysis and denaturation experiments on the minibody were fully consistent with those expected from a compact, all-β protein. For the most stable variant (MB1-K$_3$), the solubility of the monomeric form was 350 μM at neutral pH. We demonstrated that the CD spectrum was not affected by protein concentration over a range of 10–350 μM, whereas aggregation takes place above this concentration [12]. Further evidence of the minibody fold was revealed by the successful engineering of a metal-binding site onto the molecule. The grafting was accomplished by introducing three histidine residues at appropriate sites, one in H1 and two in H2 and the resulting molecule exhibited metal selectivity and relative affinities of those observed in carbonic anhydrase B, the template protein used for the metal-binding site grafting [9]. Both the metal-free and metal-bound forms of the minibody are compact, globular monomers in solution. The stability of the minibody while undergoing denaturation was measured by monitoring the mean residue ellipticity at 217 nm as a function of the urea concentration. The protein unfolds with a co-operative transition: its $\Delta G_D^{H_2O}$ value of 2.5 kcal/mol, although lower than the values (3–15 kcal/mol) observed for naturally occurring proteins [13], is similar to that of the reduced immunoglobulin C_L domain (1.8 kcal/mol) [14]. It is worthy of note that the minibody does not contain the disulfide cross-link typical of the immunoglobulin fold. The tertiary structural properties of the minibody and its variants were analyzed by near-UV CD spectroscopy. The asymmetry associated with the minibody aromatic residues results in a negative CD signal in the aromatic region of the spectrum [12]. The presence of bands in this region of the spectrum is a clear indication of the existence of a tertiary structure contrary to what could be expected for the so-called 'molten globule' folding intermediates [15]. The lack of binding to ANS at a protein concentration below 30 μM, together with the results of tryptophane fluorescence experiments carried out at neutral pH, indicate that the minibody is a more native-like species than molten globules.

Unfortunately, the millimolar concentration levels necessary for NMR studies could be achieved only under strong acidic pH (2.8). The spectroscopic analyses performed under these conditions revealed the presence of about 50 long-range (NOE) interactions and several slow exchanging amide protons that would indicate the presence of a relatively stable hydrophobic core even at this extremely acidic pH. However, the number of such long-range NOE interactions was not sufficient for a detailed structural characterization [Barbato et al., unpubl. results]. The result obtained by both NMR and limited proteolysis experiments [unpubl. results] are being used as a guide to improve our designed framework. Principally, we are aiming to further improve both solubility and thermodynamic stability of minibodies to allow a thoroughly three-dimensional characterization. Currently, we have produced two promising framework mutants, one with a more cooperative denaturation transition and a second with a stabilizing disulfide cross-linkage. Both molecules are presently being analyzed by NMR spectroscopy. An alternative approach is also being pursued to generate 'improved' minibody frameworks: that of a genetic screening for mutants with improved solubility and metabolic stability by exploiting a format that allows minibody molecules to be expressed in the periplasmic space of *E. coli*. We are characterizing some of these variants to establish whether their enhanced metabolic stability correlates with an improved thermodynamic stability.

Taken together, the biochemical studies of the minibody supported the model and suggested that its β-sheet framework could effectively serve as scaffold for engineering functional centres either by rational design [9] or by phage display [16].

Phage Display: In vitro Molecular Evolution

In phage display technology, peptide or protein domains cloned as fusions to the coat proteins (pIII and pVIII) of filamentous bacteriophage f1 are displayed on the capsid that encloses the viral genome. Mutant proteins with new biological activities can be randomly designed by comprehensive combinatorial randomization of key residues, and these proteins can be displayed on the phage surface. Proteins of interest and their associated phage can be selected from such large libraries by affinity purification using an appropriate ligate, and amplified by passage through a bacterial host. Subsequent rounds of selection and amplification result in the enrichment of the relevant phage in the mutant pool, so that individual clones can finally be isolated, and the displayed peptide is identified from sequencing the relevant portion of the phage genome [17].

Protein-protein interactions, which involve complicated networks of contact, can be rapidly investigated using this technology because large numbers of combinatorial variants can be screened simultaneously. Strategies for selecting phages that express polypeptide mutants have been applied to many different ligand/ligate pairs. Phage display is also a means of improving on nature: mutants of growth hormone, CNTF and neutrophil elastase inhibitor have been constructed whose affinities for their receptors range from 10- to 100-fold higher than the unmutagenized natural ligands [18–20]. Repertoires of small peptides have been displayed on phage and screened both with antibodies and nonantibody molecules, leading to the identification of new ligands that do not necessarily resemble their natural counterparts, yet display similar binding specificity. Large collections of antibody fragments from immunized (reviewed by Parren and Burton, [21], this issue) or naive sources [reviewed in ref. 2] have been expressed on phage and successfully screened with various antigens. The binding affinities for the rescued molecules have ranged from micromolar to nanomolar, depending on the library size; as a rule of thumb, the bigger the library, the better the binding affinity.

A variety of strategies have been developed to increase the affinity of antibody fragments for their target antigens, i.e., error-prone PCR mutagenesis

[23], chain shuffling [24], codon-based mutagenesis [25], and CDR walking [26]. The overall complexity of a library, mainly limited by bacterial transformation efficiency, has been improved dramatically by using the lox P/Cre system to exploit in vivo recombination between two distinct libraries, one for each antibody chain. Using this strategy, a combinatorial repertoire approaching 10^{11} members has been generated and sampled successfully, yielding a large panel of antibody fragments specific for many different antigens with affinities in the nanomolar range or lower [27]. Other exciting developments are in progress in several laboratories concerned with establishing mechanism-based selection schemes for the generation of catalytic activities [reviewed in ref. 5].

Minibody Repertoire

On the basis of the biochemical characterization of the minibody, we reasoned that the design of our framework was satisfactory, and that the molecule's properties justified the construction of a repertoire of minibodies on phage displaying a large number (possibly hundreds of millions) of different loop sequences. We were then in a position to establish the extent to which the surface of our small polypeptide could be modified to generate affinity-selectable ligands with reasonable affinity and specificity for a given target. As a first step, we constructed a library of about 50 million minibodies fused to the N-terminus of pIII protein of *fdtet* phage with variegated surface residues [16, 28]. The library was engineered by extensive randomization of the regions corresponding to hypervariable loops H1 and H2 with the CDR-like sequences of the minibody.

Affinity-Selection of a Minibody Antagonist of Human IL-6

To test whether the minibody library was a good source of specific ligands, we used human interleukin-6 (huIL-6) as the target molecule. Our premise for using huIL-6 was its peculiar mode of action that makes it amenable to pharmacological intervention. IL-6 is a multifunctional cytokine whose biological activity is mediated by the sequential assembly of a complex formed by a specific membrane bound IL-6 receptor (IL-6Rα) and the signal-transduc-ing subunit gp130 [29]. It has been proposed that an IL-6 antagonist might be used as a therapeutic agent in some pathologies [30, 31]. Minibodies capable of binding specifically to huIL-6 were selected using a solid-phase affinity matrix, with huIL-6 covalently bound to alkylamine beads. After multiple rounds of selection, we purified and analyzed a large number of phages inter-

Table 1. Amino acid sequences of the H1 and H2 loops of affinity selected minibodies IL-6 antagonists [28, 40], and parental molecule (MB1)

Clone	H1	H2
MB1	GFTFSDF	NKGNKY
MB02	GFTFSDF	GKEEVD
MBk	RLRFSWN	GKEEVD

The central phenylalanine (F) reside in the middle of H1 region is an invariant position in the library.

acting with the immobilized huIL-6. The polypeptide encoded by one of these clones, MB02 (table 1), that displayed an H2 region with a remarkable similarity to a short stretch of amino acids of the mouse IL-6Rα, was studied in detail. Notably, the same amino acid region had been shown to be involved in binding to the cytokine [28, and references therein]. The purified MB02 was shown by several criteria to specifically bind huIL-6 with a K_D of 1–2 μM. Furthermore, it was demonstrated that MB02 does not recognize either human oncostatin M or human ciliary neurotrophic factor, two growth factors structurally related to IL-6. The secondary structure of the affinity-selected minibody, monitored by CD spectroscopy, confirmed that the conformation of the affinity-selected polypeptide was similar to that of the parental molecule. Because of a striking sequence similarity with the IL6 receptor, we wanted to test whether MB02 could act as a receptor antagonist. To this end, an in vitro receptor-binding competition assay was carried out to determine the ability of MB02 to inhibit huIL-6 binding to huIL-6-Rα, immobilized on solid phase. The result of this experiment showed that MB02 inhibits in vitro binding of huIL6-6 to huIL6-Rα in a dose-depending manner with an IC_{50} of 2.5 μM. A synthetic linear peptide corresponding to the H2 sequence of MB02 (RGKEEVD, where R is a framework residue) did not interfere with receptor binding even at a concentration of 500 μM, indicating that the activity of MB02 was not solely dependent on the sequence of the H2 region. In virtue of the fact that huIL-6Rα associates with signal transducer molecule gp130 only upon binding to huIL-6, and that gp130 does not show any binding activity for huIL-6 alone [29], we proved by an immunoprecipitation assay that MB02 also interferes with the formation of the ternary complex cytokine/receptor/gp130 in vitro. Likewise we tested the capacity of MB02 to inhibit huIL-6 biological activity in cell culture experiments. Treatment of human hepatoma Hep3B cells with huIL-6 results in the specific phosphorylation of the acute-phase response DNA binding factor, APRF. This event is easily detected through a gel shift assay using a specific oligonucleotide probe. The treatment of Hep3B cells with 10 µM MB02 produced more than 50% inhibition of huIL-6-mediated APRF activation, supporting the evidence that MB02 binds to huIL-6 free in solution and in turn impairs the formation of the ternary complex huIL-6/huIL-6Rα/gp130. Probably, by acting as a surrogate receptor, MB02 is capable of binding with the cytokine and effectively leading to inhibition of huIL-6 biological activity [28].

These results showed that it was possible to successfully combine the rational design and combinatorial approaches to construct small β-proteins with novel and useful biological functions. The potency of MB02, albeit lower than that of huIL-6 neutralizing monoclonal antibodies obtained from hyper-immunized mice [32], is well within the range of that of antibody fragments selected from 'naive' repertoires of comparable sizes [22]. The idea that single variable domains could produce binding surfaces with adequate binding constant values was suggested by an early report on the production of functional V_H domain antibodies in *E. coli* [33], and successively confirmed by the discovery that camels can mount effective immune responses using antibodies consisting solely of H chains [34, 35]. Further confirmation was provided by later work showing that it was possible to enhance the solubility of human V_H domains by mutagenizing the interface that is usually buried with the equivalent surface of the V_L domain, so as to mimic camel antibodies (*camelized antibodies*) [36]. The same group subsequently showed that synthetic repertoires of camelized V_H domains can be a good source of specific binders with affinities in the low micromolar range [37], and amenable to further improvements.

The ability to display discontinuous structured epitopes is a major advantage of the minibody over other small polypeptide scaffolds. In addition, it offers the possibility to increase the potency and selectivity of the molecules by loop swapping, as for antibody fragments. The size of minibody variable regions permits an exhaustive collection of all possible amino acid permutations for each loop ($20^6 = 64,000,000$) to be sampled, well within the reach of phage-display libraries. Being able to swap comprehensive sets of mutants allows a larger area of the surface of the protein to be optimized, hence overcoming the limits posed by *E. coli* transformation efficiency.

Recent findings, however, have suggested that ex vivo enhancement of antibody affinity by random mutagenesis may lead to a reduction in specificity that could compromise their efficacy in the various applications [38]. This potential limitation could be more dramatic when dealing with a designed polypeptide framework whose surface is smaller than that of antibody binding sites. We wanted to verify to what extent it was possible to enhance the biological activity (both in terms of binding affinity and specificity) of minibodies. To this end, we undertook the optimization of MB02 binding affinity toward huIL-6, while minimizing cross-reactivity with molecules structurally related to the target. For this purpose, we established stringent affinity-selection conditions by taking advantage of some properties of the target molecule (fig. 2). We proved that both the affinity and the specificity of the minibody for its target can be easily enhanced, albeit modestly, by rapid sequence optimization [40]. The resulting optimized minibody (MBk) is a more potent (5- to 10-fold) and specific huIL-6 antagonist than the parental MB02 both in vitro and in tissue culture assays, as shown in figure 3. By using the IAsys™ biosensor technology, we estimated a K_D of binding to huIL-6 of $2.2 \times 10^{-7}\ M$. Notably, the secondary structure content of MBk is β-type, despite both loops (accounting for 20% of sequence diversity) being different from that of the wild-type minibody. These results support the view that contrary to the framework [12, and other unpublished data], the H1 and H2 loops are tolerant to the sequence modification. In addition, by using mutants of huIL-6, we were able to define the cytokine binding site recognized by MBk as encompassing the C-terminal region of the IL-6 putative D-helix, also part of the huIL-6Rα recognition site. This finding proves, at the molecular level, the antagonistic property of the surface-optimized minibody, and emphasizes that by establishing careful experimental conditions it is possible to rescue ligands with predefined specificity and to hence overcome the potential drawbacks of this technology suggested by the aforementioned study [38].

Selection of Minibodies with Serine Protease Inhibitor Activity

We have established a recent collaboration with Dr. DeFrancesco's group at IRBM, focusing on the hepatitis C virus (HCV)-encoded NS3 protease as a candidate target for antiviral therapy. The HCV is the causative agent of non-A, non-B hepatitis [41, 42]. The viral particles contain a positive stranded RNA genome of 9.5 kB with a single open reading frame encoding for a polyprotein of 3,010–3,033 amino acids [43–46]. Upon translation, this polyprotein is processed into 9 different polypeptides that are encoded by the viral RNA genome. Several groups [47, and references therein] have shown that all proteolytic cleavages downstream of NS3 are catalyzed by a serine protease contained within the N-terminal region of NS3. The NS3 protein is a multidomain protein of 70 kD that, in addition to the protease domain at the N-terminus, contains a putative RNA-helicase at its C-terminus. Homology modelling of the active site of the NS3 protease allowed the preference for the substrate specificity residue (cysteine) at the P1 to be predicted [48]. Truncation

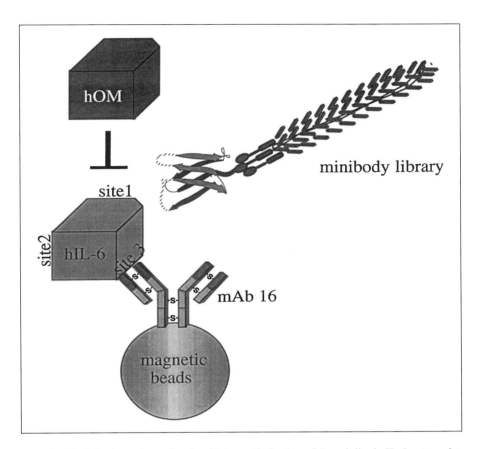

Fig. 2. Selection scheme for the affinity optimization of the minibody IL-6 antagonist. The rationale for such a selection procedure is two-fold: first, we exploited the fact that IL-6 embodies at least three distinct receptor binding sites [39]. Site 1 is the region of interaction with IL-6Rα. Site 2 is presumed to be the site of interaction with one subunit of gp130. Site 3 is the site of interaction for a second molecule of gp130 and for the mAb16. Magnetic beads coated with mAb16 were used to immobilize huIL-6 in a spatially oriented fashion and in a native conformation with respect to sites 1 and 2. This experimental scheme also has the advantage of 'masking' one of the gp130 binding sites (site 3), hence focusing the selection of ligands toward the accessible surface of the target molecule. Second, we used human oncostatin M, a cytokine structurally related to IL-6, reasoning that an excess of cytokine during the selection process would counterselect for those minibodies cross-reactive with oncostatin M and possibly related cytokines [40].

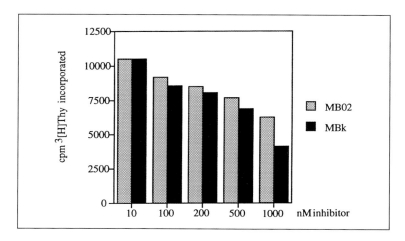

Fig. 3. Inhibition of huIL-6 biological activity in cell culture. huIL-6 growth-dependent BAF cells were treated with increasing concentration of minibodies for 20 h. The cell-associated [³H]thymidine (incorporated counts per minute) are plotted on the Y-axis (average of triplicate samples) as a function of inhibitor concentration.

experiments at both C- and N-termini have demonstrated that a 20-kD N-terminal fragment of NS3 is capable of performing all cleavages in both in vitro translation and transfection experiments with an efficiency indistinguishable from that of the wild-type enzyme [49].

We were interested in learning whether it was possible to rescue some variants that would bind NS3 from our synthetic minibody repertoire, and among those some that would also inhibit serine protease activity. We therefore developed a strategy to affinity-select minibodies from the library using purified NS3 protein produced in *E. coli*. Among several binders that have been partially characterized, we have identified NS3 protease inhibitors, some displaying a noncompetitive behavior, whereas one proved to be a competitive inhibitor with respect to the substrate. This protease inhibitor is of a modest potency with a K_i of 2–4 μM, yet it shows good selectivity as it is inactive on elastase and kallikrein [unpubl. results]. Studies are in progress to understand the mechanism by which these minibodies inhibit HCV protease and to improve their potency.

Outlook

Minibody-derived polypeptides could have special applications as biological tools and possibly also for those in vivo applications where small size may be critical for tissue penetration and short half-life for rapid serum clearance. A recent report describing the use of a pentadecapeptide derived from the CDR3 of a tumor-associated monoclonal antibody for tumor imaging highlights one possible clinical application for such reagents [50]. We can foresee the utility of designer molecules such as the minibody also in gene therapy, more specifically as targeting elements, or as intracellular inhibitors [51]. This latter application, if not a definite therapy, would at least provide a *proof of principle* for validating drug targets amenable to more traditional approaches that make use of small-molecular-weight compounds. The limit to which it is possible to optimize both affinity and specificity of minibodies remains to be established. Natural small polypeptide domains have been used by other research groups as frameworks with a similar intent, and show promising potential [52–55].

In virtue of the prediction that residues containing pharmacophoric information are predominant within protein turns [56], another potential application of minibody-like polypeptides may be that of providing template molecules (CDR-mimetic leads) that could serve as a basis for medicinal chemistry development. With a few exceptions, the clinical applications of biopharmaceuticals have till now been severely limited. The drawbacks of polypeptide therapeutics are their poor bioavailability, rapid proteolytic degradation and clearance, and immunogenicity. Converting bioactive peptides to acceptable pharmaceutical entities and understanding how to accomplish all the steps in between opens up endless possibilities for the conversion of a prolific source of lead molecules (through phage display) into potential drug candidates.

If the objective is that of interfering with large protein-protein interfaces, it is worth making a few considerations. Upon formation of the complex with the cognate protein epitope, antibodies usually appear to bury quite a large surface (approximately $1,500\ \text{Å}^2$). Their paratope consists of approximately 15–20 amino acid residues, as shown by the three-dimensional structures determined to date [6]. Clearly, if all these residues on the protein binding sites make equal contributions to the binding energy between the two proteins, then the task of mimicking the protein-binding regions with small organic molecules would obviously be difficult. However, experimental evidence suggests that, on average, four residues are sufficient to define an antigenic epitope [17, and references therein]. Likewise, synthetic peptides derived from amino acid sequences of antibody CDRs may have binding properties similar to those of the intact antibody [57]. Cunningham and Wells [58] and Clackson and Wells [59], by means of a comprehensive analysis of the energetic importance of the 31 side chains buried at the interface between human growth hormone and the extracellular domain of its receptor, have provided evidence that the functional binding epitope is much smaller than the structural epitope (determined by crystallography). In addition, theoretical calculations suggest that 'protein antigenicity involves active, attractive contributions mediated by a few energetic amino acids and a passive surface complementarity contributed by the surrounding surface area' [60].

Taken together, these elements suggest that it may be possible to mimic functional protein sites with low-moleculer-weight compounds. Initially, it is necessary to identify the key amino acid residues on the protein binding sites, followed by the design and synthesis of compounds that mimic the receptor-bound conformation of these key amino acid residues [61]. If we wish to accomplish the objective of designing small compounds that mimic the functional sites of a protein at an acceptable rate of success, it becomes crucial to dissect large proteins into smaller, conformationally restricted components that retain the bioactivity of the native protein. To this end, the minibody format may simplify the identification and characterization of the 'minimal recognition unit' responsible for biological activity. In addition, because the backbone conformations (canonical strutures) of antibody loops can be predicted, strategies for designing effective first-generation peptidomimetics based on these loops have been established [for reviews, see ref. 62, 63]. Recent data have shown that an antibody mimic (whose design was based on the crystal structure of the CDR3 of an anti-influenza N9 sialidase), containing only 4 of the 17 or so residues of the antibody that form contacts with the antigen, binds approximately 3 orders of magnitude less than that of the whole protein [64]. Considering the large difference in bound surface area between the antibody-antigen complex and that of the antibody mimetic-antigen, it was to be expected that the mimetic would bind with a lower affinity than the parent protein simply due to the difference in bound surface area (and hence a potential loss of enthalpic and hydrophobic stabilization). However, it should be possible to improve the affinity of these compounds by further chemical development. The merging of protein design tools with traditional medicinal and combinatorial chemistries carries the potential for transforming the way we currently think about the making of new pharmacological agents.

Acknowledgments

We are profoundly indebted to our colleagues A. Tramontano, E, Bianchi, A. Pessi, C. Toniatti, A.L. Salvati, G. Ciliberto, C. Steinkuehler, R. Defrancesco, S. Altamura, G. Barbato, R. Bazzo, C. Volpari, O. Epifano, R. Cortese, whose contributions to the various aspects of this project have been invaluable. We are also grateful to B. McManus and J. Clench for editing the manuscript.

References

1   Richardson JS, Richardson CD, Tweedy NB, Gernert KM, Quinn TP, Hecht MH, Erickson BW, Yan Y, McClain RD, Donlan ME, Surle MC: Looking at proteins: Representation, folding, packing, and design. Biophys J 1992;63:4175–4184.

2   Betz SF, Bryson JW, DeGrado WF: Native-like and structurally characterized designed α-helical bundles. Curr Opin Struct Biol 1995;5:457–463.

3   Kamtekar S, Schiffer JM, Xiong H, Babik JM, Hecht MH: Protein design by binary patterning of polar and nonpolar amino acids. Science 1993;262:1680–1685.

4   Sander C: Design of protein structures: Helix bundles and beyond. Trends Biotechnol 1994;12:163–167.

5   O'Neil KT, Hoess RH: Phage display: Protein engineering by directed evolution. Curr Opin Struct Biol 1995;5:443–449.

6   Wilson IA, Stanfield RL: Antibody-antigen interactions: New structures and new conformational changes. Curr Opin Struct Biol 1994;4:857–867.

7   Chothia C, Lesk AM, Gherardi E, Tomlinson IM, Walter G, Marks JD, Llewelyn MB, Winter G: Structural repertoire of the human V_H segments. J Mol Biol 1992;227:799–817.

8   Kubiak T, Whitney DB, Merrifield RB: Synthetic peptides V_H(27–68) and V_H(16–68) of the myeloma immunoglobulin M603 heavy chain and their association with the natural light chain to form an antigen binding site. Biochemistry 1987;6:7849–7855.

9   Pessi A, Bianchi E, Crameri A, Venturini S, Tramontano A, Sollazzo M: A designed metal-binding protein with a novel fold. Nature 1993;362:367–369.

10 Tramontano A, Bianchi E, Venturini S, Martin F, Pessi A, Sollazzo M: The making of the minibody: An engineered β-protein for the display of conformationally-constrained peptides. J Mol Recogn 1994;7:9–24.

11 Bianchi E, Sollazzo M, Tramontano A, Pessi A: Chemical synthesis of a designed β-protein through the flow-polyamide method. Int J Pept Protein Res 1993;41:385–393.

12 Bianchi E, Venturini S, Pessi A, Tramontano A, Sollazzo M: High level expression and rational mutagenesis of a designed protein, the minibody: From an insoluble to a soluble molecule. J Mol Biol 1994;236:649–659.

13 Creighton TE: Protein folding. Biochem J 1990;270:1–16.

14 Goto Y, Hamagouchi K: The role of intrachain disulfide bond in the conformation and stability of the constant fragment of immunoglobulin light chain. J Biochem 1979;86:1433–1441.

15 Dobson CM, Evans PA, Radford SE: Understanding how proteins fold: The lysozyme story so far. Trends Biochem Sci 1994;19:31–37.

16 Venturini S, Martin F, Sollazzo M: Phage display of the minibody: A β-scaffold for the selection of conformationally-constrained peptides. Protein Pept Lett 1994;1:70–75.

17 Cortese R, Monaci P, Nicosia A, Luzzago A, Felici F, Galfrè G, Pessi A, Tramontano A, Sollazzo M: Identification of biologically active peptides using random libraries displayed on phage. Curr Opin Biotech 1995;6:73–80.

18 Roberts BL, Markland W, Ley AC, Kent RB, White DW, Guterman SK, Ladner RC: Directed evolution of a protein: Selection of potent neutrophil elastase inhibitors displayed on M13 fusion phage. Proc Natl Acad Sci USA 1992;89:2429–2433.

19 Lowman HB, Wells JA: Affinity maturation of human growth hormone by monovalent phage display. J Mol Biol 1993;234:564–578.

20 Saggio I, Cloaguen I, Poiana G, Laufer R: CNTF variants with increased biological potency and receptor selectivity define a functional site of receptor interaction. EMBO J 1995;14:3045–3054.

21 Parren PWHI, Burton DR: Antibodies against HIV-1 from phage display libraries: Mapping of an immune response and progress towards antiviral immunotherapy; in Capra JD (ed): Antibody Engineering. Chem Immunol. Basel, Karger, 1997, vol 65, pp 18–56.

22 Winter G, Griffiths AD, Hawkins RE, Hoogenboom HR: Making antibodies by phage display technology. Annu Rev Immunol 1994;12:433–455.

23 Gram H, Marconi L-A, Barbas CF III, Collet TA, Lerner RA, Kang AS: In vitro selection and affinity maturation of antibodies from a naive combinatorial immunoglobulin library. Proc Natl Acad Sci USA 1992;89:3576–3580.

24   Marks JD, Griffiths AD, Malmqvist M, Clackson TP, Bye JM, Winter G: Bypassing immunization: Building high affinity human antibodies by chain shuffling. Bio/Technology 1992;10:779–783.

25   Glaser SM, Yelton DE, Huse WD: Antibody engineering by codon-based mutagenesis in a filamentous phage vector system. J Immunol 1992;149:3903–3913.

26   Yang W-P, Green K, Pinz-Sweeney S, Briones AT, Burton DR, Barbas CF III: CDR walking mutagenesis for the affinity maturation of a potent human anti HIV-1 antibody into the picomolar range. J Mol Biol 1995;254:392–403.

27   Griffiths AD, Williams SC, Hartley O, Tomlison IM, Waterhouse P, Crosby WL, Kontermann RE, Jones PT, Low NM, Allison TJ, Prospero TD, Hoogenboom HR, Nissim H, Cox JPL, Harrison JL, Zaccolo M, Gherardi E, Winter G: Isolation of high affinity human antibodies directly from large synthetic repertoires. EMBO J 1994;13:3245–3260.

28   Martin F, Toniatti C, Salvati AL, Venturini S, Ciliberto G, Cortese R, Sollazzo M: The affinity-selection of a minibody polypeptide inhibitor of human interleukin-6. EMBO J 1994;13:5303–5309.

29   Kishimoto T, Taga T, Akira S: Cytokine signal transduction. Cell 1994;76: 253–262.

30   Akira S, Taga T, Kishimoto T: Interleukin-6 in biology and medicine. Adv Immunol 1994;54: 1–78.

31   Savino R, Lahm A, Salvati AL, Ciapponi L, Sporeno E, Altamura S, Paonessa G, Toniatti C, Ciliberto G: Generation of interleukin-6 receptor antagonists by molecular-modeling guided mutagenesis of residues important for gp130 activation. EMBO J 1994;13:1357–1367.

32   Brakenhoff JPJ, Hart M, De Groot ER, Di Padova F, Aarden LA: Structure-function analysis of human IL-6. J Immunol 1990;145:561–568.

33   Ward SE, Güssow D, Griffiths AD, Jones PT, Winter G: Binding activities of a repertoire of single immunoglobulin variable domains secreted from *Escherichia coli.* Nature 1989;341:544–546.

34   Hamers-Casterman C, Atarhouch T, Muyldermans S, Robinson G, Hamers C, Bajyana Songa E, Bendahaman N, Hamers R: Naturally occurring antibodies devoid of light chains. Nature 1993; 363:446–448.

35   Muyldermans S, Atarhouch T, Saldanha J, Barbosa JARG, Hamers R: Sequence and structure of V_H domain from naturally occurring camel heavy chain immunoglobulins lacking light chains. Protein Eng 1994;7:1129–1135.

36   Davies J, Riechmann L: 'Camelising' human antibody fragments: NMR studies on V_H domains. FEBS Lett 1994;339:285–290.

37   Davies J, Riechmann L: Antibody V_H domains as small recognition units. Bio/Technology 1995; 13:475–479.

38   Casson LP, Manser T: Random mutagenesis of two complementarity determining region amino acids yields an unexpectedly high frequency of antibodies with increased affinity for both cognate antigen and autoantigen. J Exp Med 1995;182:743–750.

39   Paonessa G, Graziani R, De Serio A, Savino R, Ciapponi L, Lahm A, Salvati AL, Toniatti C, Ciliberto G: Two distinct and independent sites on IL-6 trigger gp130 dimer formation and signalling. EMBO J 1995;14:1942–1951.

40   Martin F, Toniatti C, Salvati AL, Ciliberto G, Cortese R, Sollazzo M: Coupling protein design and in vitro selection strategies: Improving specificity and affinity of a designed β-protein IL-6 antagonist. J Mol Biol 1996;255:86–97.

41   Choo QL, Kuo G, Weiner AJ, Overby LR, Bradley DW, Houghton M: Isolation of a cDNA clone derived from a blood-borne non-A, non-B viral hepatitis genome. Science 1989;244:359–362.

42   Kuo G, Choo QL, Alter HJ, Gitnick GL, Redecker AG, Purcell RH, Myamura T, Dienstag JL, Alter MJ, Syevens CE, Tagtmeyer GE, Bonino F, Colombo M, Lee WS, Kuo C, Berger K, Shister JR, Overby LR, Bradley DW, Houghton M: An assay for circulating antibodies to a major etiologic virus of human non-A, non-B hepatitis. Science 1989;244:362–364.

43   Kato M, Hijikata M, Ootsuyama Y, Nakagawa M, Ohkoshi S, Sugimura T, Shimotohno K: Molecular cloning of human hepatitis C virus genome from Japanese patients with non-A, non-B hepatitis. Proc Natl Acad Sci USA 1990;87:9524–9528.

44   Choo QL, Richman KH, Ham JH, Berger K, Lee C, Dong C, Gallegos C, Coit D, Medina-Selby A, Barr PJ, Weiner AJ, Bradley DW, Kuo G, Houghton M: Genetic organization and diversity of the hepatitis C virus. Proc Natl Acad Sci USA 1991;88:2451–2455.

45 Takamizawa A, Mori C, Fuke I, Manabe S, Murakami S, Fujita J, Onoshi E, Andoh T, Yoshida I, Okayama H: Structure and organization of the hepatitis C virus genome isolated from human carriers. J Virol 1991;65:1105–1113.
46 Grakoui A, Wychowski C, Lin C, Feinstone SM, Rice CM: Expression and identification of hepatitis C virus polyprotein cleavage products. J Virol 1993;67:1385–1395.
47 Tomei L, Failla C, Santolini E, De Francesco R, LaMonica N: NS3 is a serine protease required for processing of hepatitis C virus polyprotein. J Virol 1993;67:4017–4026.
48 Pizzi E, Tramontano A, Tomei L, LaMonica N, Failla C, Sardana M, Wood T, DeFrancesco R: Molecular model of the specificity pocket of the hepatitis C virus protease: Implication for substrate recognition. Proc Natl Acad Sci USA 1994;91:888–892.
49 Failla C, Tomei L, DeFrancesco R: An amino-terminal domain of the hepatitis C virus NS3 protease is essential for interaction with NS4A. J Virol 1995;69:1769–1777.
50 Sivolapenko GB, Douli V, Pectasides D, Skarlos D, Sirmalis G, Hussain R, Cook J, Courtenay-Luck NS, Merkouri E, Konstantinides K, Epenetos AA: Breast cancer imaging with radiolabelled peptide from complementarity-determining region of antitumor antibody. Lancet 1995;346:1662–1666.
51 Duan L, Zhu M, Bagasra O, Pomerantz RJ: Intracellular immunization against HIV-1 infection of human T lymphocytes: Utility of anti-rev single-chain variable fragments. Hum Gene Ther 1995;6:1561–1573.
52 Bianchi E, Folgori A, Wallace A, Nicotra A, Acali S, Phalipon A, Barbato G, Bazzo R, Cortese R, Felici F, Pessi A: A conformationally homogeneous combinatorial peptide library. J Mol Biol 1995;247:154–160.
53 Ku J, Schultz PG: Alternate protein frameworks for molecular recognition. Proc Natl Acad Sci USA 1995;92:6552–6556.
54 McConnell SJ, Hoess RH: Tendamistat as a scaffold for conformationally constrained phage peptide libraries. J Mol Biol 1995; 250:460–470.
55 Nord K, Nilsson J, Nilsson B, Uhlen M, Nygren P-A: A combinatorial library of an α-helical bacterial receptor domain. Protein Eng 1995;8:601–608.
56 Rose GD, Gierasch LM, Smith JA: Turns in peptides and proteins. Adv Protein Chem 1985;37:1–109.
57 Taub R, Greene MI: Functional validation of ligand mimicry by anti-receptor antibodies: Structural and therapeutic implications. Biochemistry 1992;31:7431–7435.
58 Cunningham BC, Wells JA: Comparison of a structural and functional epitope. J Mol Biol 1993; 234:554–563.
59 Clackson T, Wells JA: A hot spot of binding energy in a hormone-receptor interface. Science 1995; 267:383–386.
60 Novotny J, Bruccoleri RE, Saul FA: On the attribution of binding energy in antigen-antibody complexes McPC603, D1.3, and HyHEL-5. Biochemistry 1989;28:4735–4746.
61 Hruby VJ, Al-Obeidi F, Kazmierski W: Emerging approaches in the molecular design of receptor-selective peptide ligands: Conformational topographical and dynamic consideration. Biochem J 1990;268:249–262.
62 Sollazzo M, Bianchi E, Felici F, Cortese R, Pessi A: Conformationally defined peptide libraries on phage: Selectable templates for the design of pharmacological agents; in Cortese R (ed): Combinatorial Libraries. Berlin, de Gruyter, 1996, pp 127–143.
63 Zhao B, Helms LR, Des Jarlais RL, Abdel-Meguid SS, Wetzel R: A paradigm for drug discovery using a conformation from the crystal structure of a presentation scaffold. Nat Struct Biol 1995; 2:1131–1137.
64 Smythe ML, von Itzstein M: Design and synthesis of a biologically active antibody mimic based on an antibody-antigen crystal structure. J Am Chem Soc 1994;116:2725–2733.

Dr. Maurizio Sollazzo, Department of Biotechnology, IRBM P. Angeletti,
Via Pontina km 30.600, I–00040 Pomezia (RM) (Italy)

Capra JD (ed): Antibody Engineering.
Chem Immunol. Basel, Karger, 1997, vol 65, pp 18–56

....

Antibodies against HIV-1 from Phage Display Libraries: Mapping of an Immune Response and Progress towards Antiviral Immunotherapy

Paul W.H.I. Parren, Dennis R. Burton

The Scripps Research Institute, Department of Immunology, La Jolla, Calif., USA

Antibodies probably were the first proteins to be recognized as potential therapeutics. In the 1890s, Behring and Kitasato [1] and Behring [2, 3] demonstrated therapeutic efficacy of immune serum against tetanus and diphtheria. Despite this early success, however, the use of passive immunotherapy has remained fairly limited. This probably reflects the great success of antibiotics in combatting bacterial diseases, of vaccines in the prevention of viral diseases, and the difficulties of obtaining antibodies useful for passive immunization. The AIDS pandemic associated with widespread opportunistic infectious agents in a large pool of immunocompromised hosts has once more focused attention on infectious disease.

Neutralizing antibodies are an important component of the host defense to viral infections, and are thought to play a major role in limiting the spread of cell-free virus, whereas killing of virally infected cells is mostly attributed to the action of cytotoxic T cells. In HIV-1 infection, passive immunotherapy could play an important role in immunoprophylaxis and immunotherapy, and understanding of neutralization epitopes is critical for the development of vaccines.

Hyperimmune human serum preparations have proved useful in a number of instances, notably to achieve a great decrease in the incidence of hemolytic disease of the newborn through anti-rhesus D prophylaxis in the developed countries [4]. In therapy, a preparation of specific pooled human sera administered within 8 days of the onset of symptoms is used to prevent death due to Argentine hemorrhagic fever [5]. The last example may seem exotic, but it is our contention that antibodies would be used in the treatment of many more infectious diseases if potent human monoclonal antibodies were available.

Immunotherapy prefers a ready supply of specific human antibodies. The disadvantages of hyperimmune serum are its limited availability and the relatively small and strongly variable proportion of specific functional antibodies in the preparation. The discovery of hybridoma technology by Köhler and Milstein [6] offered the availability of unlimited supplies of specific monoclonal antibodies (mAb). However, in vivo application of mAbs in human medicine has suffered from the drawback that most mAbs generated were of rodent origin, which induced human antirodent antibody immune responses making long-term immunotherapy difficult. Generating cell lines producing human antibodies has proved to be tedious and inefficient, in stark contrast to the relative ease in making rodent mAbs. The problems encountered are mostly technical, and include (1) poor availability of human myeloma fusion partners; (2) chromosomal instability of mouse myeloma/human B-cell hybrids; (3) low efficacy of Epstein-Barr-virus-mediated immortalization of human B-cells and poor ability of these cells to grow at low cell densities, and (4) the use of circulating peripheral blood lymphocytes as the source of human immune cells, whereas this cell population primarily contains B cells not actively involved in an ongoing immune response [7–10]. The issue is further complicated by ethical considerations with respect to immunization of humans for antibody production, and tolerance which averts immune responses directed against many potential therapeutically interesting human antigens. 'Humanization' of existing mouse mAbs by variable domain replacement [11–15] or by complementarity-determining region (CDR) grafting [16, 17] has been achieved to transfer antigen specificities from rodent to human antibody frameworks. The chimeric antibodies are easily produced but retain mouse variable domains with possible consequences for immunogenicity [18, 19]. CDR grafting produces nearly human antibodies but these are more difficult to construct and may require adjustments to human frameworks [20]. Further, one may be unable to obtain therapeutically interesting antibodies against certain human pathogens, such as for example HIV-1, from immunizations in rodents due to peculiarities of immunization in the animal model or difficulties in obtaining antigens for immunization.

Novel methodologies to tap the human immune repertoire directly at the molecular level have been developed in the last 7 years. These approaches became possible by advancement in several areas. First, through the successful production of functionally assembled antibody Fab fragments in *Escherichia coli* by directing secretion into the periplasmic space leading to disulfide bond formation and proper folding of antibody domains [21]. Second, the development of the polymerase chain reaction (PCR) allowing rapid cloning of antibody cDNA [22, 23]. These methods were quickly applied to prepare combinatorial antibody libraries in λ-phage from spleen mRNA from an immunized mouse [24] or peripheral blood lymphocyte mRNA from tetanus-boosted human donors [25, 26]. Specific antibodies were obtained by screening these libraries by incubating purified labeled antigen with plaque lifts on nitrocellulose filters. This screening procedure however restricts the size of the library which may be examined by the number of plates that can be experimentally handled and further limits one to the isolation of antibodies against antigens that are compatible to labeling and hybridization in a filter lift assay [27].

Recombinant antibody technology finally came of age when lessons learned from the λ-phage combinatorial libraries were combined with phage display. Smith [28] had already demonstrated that a linkage between phenotype and genotype could be established in filamentous bacteriophage. He showed that foreign DNA fragments could be inserted into filamentous phage gene III, which codes for the phage coat protein cpIII, to create a fusion protein with the foreign sequence in the middle. The fusion protein was incorporated into the virion, which retained infectivity and displayed the foreign peptide in a form accessible to specific antibody to the peptide. The fusion phage could be greatly enriched relative to ordinary phage by affinity selection on immobilized antibody (a process usually termed 'panning'). The concept of selectable phage display libraries was subsequently established with peptide libraries [29–31]. The expression of recombinant proteins on phage, such as antibody Fab fragments, in phage display libraries followed shortly thereafter [32–34]. Antibody phage display libraries are now routinely prepared and may completely replace conventional technology in the near future. However, how does phage display technology hold up to its promise? Are human antibodies of therapeutic significance being delivered? In order to answer these questions, we will review (1) the mapping of antiviral immune responses in HIV-1 using phage display identifying neutralizing epitopes, and (2) the progress on a therapeutically promising antibody to HIV-1.

Antibody Phage Display Libraries

Before preparing the library, it is crucial to identify suitable donors and to choose tissue source and antibody isotype. The choice of donor and isotype is dictated primarily by the serum titer against the antigen to be studied. In studying infectious disease, these are preferably persons who have cleared or controlled an infection and have high neutralizing antibody serum titers against the agent of interest. A high serum titer is presumed to reflect a vigorous humoral immune response with correspondingly high levels of specific mRNA with which to begin the cloning process. The specific mRNA concentration will be highest in a tissue rich in antibody-secreting plasma cells. It appears from studies from tetanus boosting of humans that, immediately following secondary antigen contact (i.e. 3–10 days), the number of specific antibody-secreting cells in the peripheral blood is high [35–38]. Indeed human Fabs to tetanus toxoid have been derived from peripheral blood lymphocyte libraries from boosted donors [25, 26]. However, following the antigen boost, the number of antibody-secreting cells in the periphery declines very rapidly to a low resting level [35, 37]. Consistent with this was the failure to detect antitetanus toxoid antibodies in a library prepared from an individual with a high serum titer to toxoid but no recent boost [25]. Peripheral blood lymphocytes may further be a valuable tissue source in the case of chronic antigen stimulation as for example in autoimmune disease such as systemic lupus erythematosus [39]. Bone marrow, however, has been shown to be the major depository of antibody-producing cells in humans [10] and is our tissue of choice to prepare a library. We have furthermore shown that antibodies to many different pathogens can be derived from bone marrow libraries constructed from a single donor [40].

A general outline of the procedure used to prepare antibody phage display libraries in our laboratory is shown in figure 1. Total RNA is prepared from the tissue source and mRNA is reverse-transcribed into cDNA using an oligo d(T) primer. Using different sets of primers in PCR, we then amplify the cDNA encoding the heavy-chain Fd fragment (variable and C_H1 domain) and the whole of κ and λ light chains (fig. 2). For each heavy- and light-chain PCR, one specific 3' primer is combined with several 5' primers to amplify antibody cDNA from all V_H and V_L gene families [25, 27]. Pools of V_H and V_L DNA are then cut with restriction enzymes (*XhoI/SpeI* and *XbaI/SacI*, respectively) and cloned into the phagemid vector pComb3H [34]. Using electroporation, libraries of $> 10^7$ clones can be routinely prepared. The diversity in immune libraries of such size is in our hands usually large enough to obtain a number of high-affinity antibodies against the antigen of interest if there was a detectable immune response in the serum. Note that such immune libraries are skewed towards a strong representation of high-affinity antibodies which were being produced in the active immune response at the time of tissue donation.

The library of antibody cDNAs in the *E. coli* culture after transformation is then expressed on bacteriophage by superinfection with helper phage. Mature phage extruded from the *E. coli* cells contain the pComb3H phagemid ssDNA with the cloned antibody fragments packaged in the phage protein coat. Both native cpIII protein (necessary for infection) and Fab-cpIII fusion protein (typically in a single copy per phage) are present on one of the extreme ends of the phage filament. We refer to Barbas and Burton [27] for a more detailed discussion on antibody phage display vector systems.

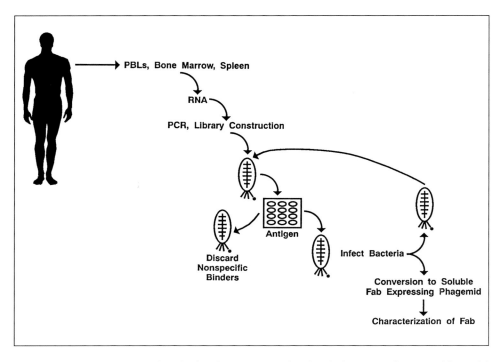

Fig. 1. Strategy for cloning human monoclonal Fab fragments from combinatorial libraries by phage display. RNA prepared from antibody-producing tissue sources from an immunized donor is reverse transcribed, and the light chain and Fd portion of the heavy chain are amplified using PCR. The amplified genes are then cloned sequentially into a phagemid vector and 'rescued' to a phage display library in which each phage expresses Fab on its surface. Specific Fab-phage are selected by virtue of binding to antigen immobilized on a surface such as an ELISA well, and converted to a soluble Fab-expressing system for further characterization. After Burton and Barbas [155].

The phage display library is then taken through several rounds of panning to select for antigen-specific clones (fig. 1). Antigen immobilized (e.g. on a plastic surface) is incubated with the phage display library to allow specific phage-Fabs to bind, after which nonbound phage-Fabs are removed by vigorous washing. Bound phage-Fabs are then eluted (e.g. low pH or excess soluble antigen) and amplified by infection of a fresh *E. coli* culture, and the above process is repeated. The number of different phage-Fabs in the mixture will rapidly diminish, focusing in on specific phage-Fabs binding with the highest affinity. Finally, phagemid DNA is prepared from the remaining phage-Fabs after infecting another *E. coli* culture, and reconstructed by excision of the gene fragment encoding cpIII. Reconstructed phagemid can then be used to

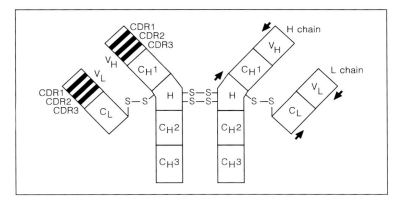

Fig. 2. The human IgG1 molecule. For generation of Fab phage display libraries, PCR primers are chosen to amplify DNA corresponding to the Fd part ($V_H C_H 1$) of the heavy chain and the whole of the light chain. The positions of the primers are represented by arrows at equivalent positions on the protein. The 3' primer for the heavy chain hybridizes to a hinge region sequence to include the cysteine involved in the heavy-light chain disulfide bond in IgG1. After Burton and Barbas [27].

transform bacteria for the production of soluble Fab fragments. At this time we usually screen for specific antibody clones by testing Fab containing *E. coli* extracts for antigen binding in an immunoassay (e.g. ELISA). For further reading on phage display of peptides, antibodies and other proteins there are many recent excellent reviews [27, 41–49].

Expression of Whole Antibody Molecules

Antibodies derived by phage display are obtained as Fab fragments. For most therapeutic applications, whole antibodies are preferable to Fab fragments. The most important being antibody in vivo half-life, which is extremely short for Fab fragments that are excreted through renal filtration due to their small size (50 kD). Expression of whole antibody molecules needs to be done in eukaryotic cells, at least because glycosylation of the Fc part at amino acid Asn297 is essential in the recognition of IgG by IgG Fc receptors [50, 51]. Cell lines of choice are myeloma [52], or Chinese hamster ovary (CHO) cells [53] although a baculovirus expression system in insect cells has also been used [54]. Modification of existing vectors enables the facile expression of whole antibodies utilizing Fab cassettes from the phage system. Using such an approach we have expressed a whole IgG1 molecule in CHO cells using the Fd and light chain derived from a phage clone [55]. Recently, we reconstructed a recombinant Fab fragment into a whole IgG1 molecule by use of a glutamine synthetase selection plasmid transfected into CHO cells [56, 57]. In glutamine-free medium, glutamate and ammonia are metabolized into glutamine by glutamine synthetase which can be inhibited by addition of the specific inhibitor methionine sulfoximine. After transfection, selection in increasing doses of methionine sulfoximine can be used to amplify the copy number of the integrated glutamine synthetase plasmid and derive subclones expressing higher levels of the recombinant antibody [57].

Antibodies to Viruses

Specific human antibodies have been shown to prevent disease caused by a wide variety of viruses belonging to diverse RNA or DNA virus families that include orthomyxoviruses, paramyxoviruses, alphaviruses, flaviviruses, arenaviruses, lentiviruses, picornaviruses, hepadnaviruses, and herpesviruses [reviewed in ref. 58].

Commercial human gamma globulin, which usually contains 16–18% IgG, of which only a very small proportion is specific for any given antigen, is highly effective in the prevention of hepatitis A disease and has been widely used for that purpose during the past 40 years [59]. Passive transfer of antibodies, in the form of immune or hyperimmune globulins, further is currently being used in the prophylaxis and treatment of a number of viral infections, including cytomegalovirus [60, 61], respiratory syncytial virus [62], hepatitis B virus [63], varicella zoster virus [64], and rabies virus [65].

In lentiviruses, most evidence for the ability of antibodies to confer a protective effect comes from studies in animal models. Protection or partial protection from HIV-1 infection by vaccination with recombinant gp120 or gp160 in chimpanzees has been described in several reports [66–70]. Passive immunization with an anti-gp1210 V3 loop mAb has also been reported to protect chimpanzees against HIV-1 infection [71]. In the macaque model, in contrast, most attempts to protect against SIV infection by envelope vaccination have been unsuccessful [72–74]. Passive transfer of hyperimmune globulin from SIV-infected animals has been shown to offer protection in two studies [74, 75], but not in another [76].

A potential therapeutic effect for antibodies in HIV-1 would be indicated by their ability to reduce viral load in infected individuals. We are not aware of any instance where this has been described convincingly although a recent study by Levy et al. [77] is suggestive.

Overall, the studies available suggest that antibodies against HIV-1 can exert an antiviral effect in vivo which, in general, however, does not seem sufficient for the sustained containment of the virus. Many studies on human mAbs against HIV-1 envelope isolated from seropositive donors imply that there is a twofold problem. First, most antibodies are relatively ineffective at neutralizing primary isolates of HIV-1 [78]. Most workers view neutralization of primary isolates as a key measure of antibody efficacy. Second, where neutralizing activity is observed, it is often relatively isolate-specific. However, as described below in detail for one antibody (b12), it is possible to potently neutralize a broad range of primary isolates. Unfortunately, neither the natural response to HIV-1 infection nor vaccination with gp160 or monomeric gp120 appear to efficiently elicit such antibodies.

Neutralization of Virus by Recombinant Fab Fragment and Whole IgG

Recombinant antibodies from phage display libraries are retrieved and first tested as Fab fragments. Only after they have shown to potently neutralize virus in vitro are they converted to whole antibody. Is this a sound strategy and does conversion affect the neutralizing ability?

Conversion of a neutralizing Fab fragment into a whole antibody will generally increase its neutralizing ability. We are unaware of any instance in the literature where an Fab fragment clearly shows stronger virus-neutralizing ability than the corresponding whole IgG antibody. There are two mechanisms which influence this difference in neutralizing ability between monovalent Fab and bivalent IgG. (1) One is of immunochemical nature: the influence of multivalency on the binding properties of the antibody. It has been shown that the two combining sites in a bivalent IgG usually do not interact to produce fully cooperative binding [79] for reasons of restricted flexibility of the bivalent antibody molecule, and flexibility or availability of the antigen. Differences in functional affinity between Fab and IgG are usually found to be between 10- and 100-fold, with the larger differences found for antihapten antibodies compared to antibodies specific for protein antigens [80–82]. (2) The second is of a biological nature: the mechanism of virus neutralization. Certain antibodies are able to neutralize virus by an antibody cross-linking event. In poliovirus, for example, antibodies have been described which neutralize the virus by aggregation, and papain digestion of the bound antibody is able to partially reverse neutralization [83–85]. Similar results have been shown for antibodies against some but not all sites in neutralization of rotavirus [86], and influenza virus [87]. In these cases one finds large differences in neutralization between monomeric Fab and bivalent IgG since the Fab cannot cross-link, and such neutralizing antibodies may be missed by using our approach. Table 1 gives a summary of studies that compare neutralization by Fab and IgG, in

Table 1. Difference between monovalent Fab and bivalent IgG in neutralization assays

Virus	Antibody source	Difference between Fab and IgG in neutralization assay	Reference
Poliovirus type I	rabbit polyclonal	2- to 3-fold	155
Poliovirus type II	rabbit polyclonal	1- to 5-fold	83
Bacteriophage R17	rabbit polyclonal	>30-fold	156
Herpes simplex virus	rabbit polyclonal	4- to 10-fold	157
Bacteriophage øX174	rabbit polyclonal	4-week immune serum: 8- to 76-fold 20-week immune serum: none	158
DNP hapten conjugated-bacterio-phage øX174	rabbit anti-DNP	100-fold	80
Paramyxovirus	rabbit anti-HN and F	similar 50% titers, Fab however does not reach 100% neutralization	159
Influenza virus	mouse monoclonal anti-HA	30- to 60-fold for group I/II Mabs, group III/IV Fabs do not neutralize	87
VEE virus	mouse monoclonal	16-fold	95
VEE virus	mouse monoclonal	10-fold	160
Rhinovirus	mouse monoclonal	13- to 61-fold	161
La Crosse virus	mouse monoclonal	4-fold	162
Rabies virus	human monoclonal, recombinant Fab, anti-G protein	10-fold	163
HIV-1	human recombinant Fab and IgG, anti-gp120	5- to 10-fold	56

instances where both did neutralize the virus. Differences in neutralization potency between Fab and IgG are found ranging from no difference to a 100-fold difference. In agreement with the immunochemical studies, the largest difference is observed in an assay where antihapten antibody and hapten-conjugated virus were used. There is remarkable agreement between the older studies using polyclonal antibody preparations, studies using mAbs and their fragments and the most recent studies using mAbs and molecularly cloned fragments. The consensus appears to be that increasing valency from monovalent to bivalent in the interaction of antibodies with complex protein antigens increases the binding and neutralization potency 5- to 50-fold.

However, does good neutralization in vitro always indicate good virus-neutralizing activity in vivo?

A good correlation between in vitro neutralization and protection has indeed been found in many studies [reviewed in ref. 88]. Exceptions have been described for murine hepatitis virus [89, 90], HSV-1 [91, 92], HSV-2 [93], and bovine coronavirus [94], where in vitro neutralizing antibody did not protect in vivo. Thus although neutralization is a useful indicator, it has limitations and the relative importance of mechanisms which contribute to antibody-mediated clearance of virus in vivo remain unclear. Some studies have suggested that an intact IgG molecule is required for efficient protection in vivo. F(ab')2 fragments of neutralizing antibodies to lymphocytic choriomeningitis virus, Venezuelan equine encephalitis virus, and yellow fever virus did not protect against infection although the corresponding whole IgG molecules did protect. Protection against lymphocytic chorio-meningitis and Venezuelan equine encephalitis virus infection with intact IgG was furthermore shown complement independent by using mice deficient for C3 or C5 [95–98]. Protection against foot-and-mouth disease virus infection required 10- to 500- fold more F(ab')2 than IgG [99], and mice challenged with Semliki forest virus were poorly or not protected by neutralizing F(ab')2 fragments [100]. In vivo protection with neutralizing F(ab')2 has been demonstrated, however, in cotton rats challenged with respiratory syncytial virus (RSV) [101]. Explanations for differences in protection between intact IgG and F(ab')2 fragments are the activation of Fc-mediated effector functions by the whole antibody and/or a more rapid catabolism of F(ab')2 fragment [102].

Some studies have shown a protective ability of neutralizing Fab fragments. Anti-RSV antibody Fab fragments were therapeutically effective for RSV-infected mice [103]. We have furthermore demonstrated some protection against infection with HIV-1 in hu-PBL-SCID mice with a neutralizing Fab fragment against the gp120 CD4-binding site (CD4bs), although protection was much less efficient than with the intact antibody (see below) [104]. Apart from the lack of Fc-mediated effector function and a much shorter serum half-life of the Fab fragment, this difference could also be explained, however, by a reduction in affinity (approximately 10-fold) for the Fab.

Antibodies to HIV-1

The *env* gene of HIV-1 encodes a polypeptide of 160 kD (gp160) which is processed into two surface glycoprotein subunits: gp120 and the transmembrane subunit gp41. The native protein on the viral surface is an oligomeric complex composed of noncovalently associated gp120 and gp41 molecules (probably trimers). The gp120 molecule interacts with the HIV-1 cell surface receptor CD4.

Most of the HIV-1-neutralizing response in the serum of infected individuals is directed against the gp120 subunit of the envelope glycoprotein [105–107]. Many hundreds of mAbs to gp120 (rodent and human) have been described and used to map the surface topology of envelope glycoprotein structure [108, 109]. Neutralizing antibodies map into several distinct epitope clusters on gp120 [reviewed in ref. 109]: (1) The third hypervariable loop of gp120 (V3 loop) was long considered the principal neutralizing domain (PND) because of its immunodominance in experimentally infected or gp120-immunized animals, and V3 loop mAbs potently neutralize HIV-1 strains passaged through T-cell lines (T-cell-line-adapted virus) [106, 110–112]. (2) Antibodies to conformational epitopes that overlap the CD4bs on gp120 [56, 78, 113–115]. (3) CD4-enhanced antibodies, that bind to gp120 with enhanced affinity in the presence of CD4 [116]. (4) The V2 loop [117]. (5) A unique epitope recognized by mAb 2G12, dependent on N-linked glycans in the C2, C3, C4 and V4 domain [118]. (6) A mostly uncharacterized cluster of neutralizing epitopes is located within the V1 loop region. This epitope cluster is immunodominant in the response to gp120 in laboratory workers infected with T-cell-line-adapted strain HIV-1 IIIB [119]. No mAbs to this region have been identified.

Only one epitope outside of gp120 is currently recognized as an epitope for potent neutralizing antibodies, this epitope is located on the gp41 ectodomain and recognized by mAb 2F5 [120, 121].

Neutralization of HIV-1 by sera or mAbs is usually done in in vitro assays, which until recently were performed with stocks of HIV-1 strains grown in T-cell lines. Studies by Daar et al. [122], however indicated that primary isolates, i.e. virus stocks prepared by limited passaging through primary cells, e.g. peripheral blood mononuclear cells (PBMCs) stimulated with phytohemagglutin (PHA), were much more resistant to inactivation by soluble CD4 than T-cell-line-adapted viruses. Similar results were found for inactivation of HIV-1 by neutralizing antibodies [78], in particular anti-V3 loop antibodies. Recently, it was found that the V3 loop, the PND for T-cell-line-adapted virus, appears to be inaccessible on the native envelope of primary viruses [123].

We have constructed a total of eight IgG1 libraries from HIV-1 seropositive individuals; three are from long-term (4–7 years) asymptomatic individuals, one is from a long-term (13 years) nonprogressor, two are from individuals with exceptionally high neutralizing serum titer to T-cell-line-adapted strains of HIV-1, one is from a donor whose CD4 count recovered following *Pneumocystis carinii* infection (although this was not unambiguously documented) and one is from a gp160 (MicroGeneSys)-vaccinated individual.

0 out of 4 samples [131]. To determine whether passage of plasma virus through PHA-activated PBMC would alter neutralization sensitivity, in vitro neutralization assays were performed against virus passaged once in PHA-activated PBMC recovered from the six HIV-1 infected plasma samples. IgG1 b12 did neutralize 5 out of 6 passaged viruses, but there was no correlation in neutralization sensitivity between ex vivo and in vitro neutralization assays. The plasma samples apparently contain viruses with varying sensitivity to neutralization and the discrepancy may relate to strong selection for a virus population representing only a minor fraction from a variable virus population upon passaging through PBMC.

Protection against HIV-1 Infection in hu-PBL-SCID Mice by Induction of Passive Immunity

As shown above, IgG1 b12 potently neutralizes a wide range of primary isolates of HIV-1 in various in vitro assays. To obtain an initial evaluation of the efficacy of neutralization and protection against infection in vivo, we injected hu-PBL-reconstituted SCID mice with graded doses of neutralizing antibody b12 (either as an Fab or whole antibody preparation) and subsequently exposed the mice to HIV-1 SF2. We first assessed the pharmacokinetics of the antibody preparations and serum half-lives of 30.2 ± 1.3 h and 7.4 ± 0.7 days were found for Fab b12 and IgG1 b12, respectively [104].

PBMC (2×10^7 in 0.5 ml of PBS) were injected intraperitoneally into CB.17 SCID/SCID mice. Two weeks after reconstitution, the mice were challenged with 100 $TCID_{50}$ of HIV-1 SF2. This dose represents 10–100 minimal animal infectious doses in this model. Antibody was given as two injections; Fab were injected intraperitoneally at –1 and +1 h, and whole antibody at –7 days and –24 h relative to inoculation with the virus. The animals were killed at 2 weeks after virus challenge and HIV infection was assessed by p24 ELISA after coculture and PCR amplification as described below. The 2-week time-point for assessing protection has been studied. In experiments with HIV-1 SF2 (n = 20) using these reporter assays, no instance of virus infection has been observed that was unapparent at 2 weeks but was detected later.

Fresh peritoneal lavage cells and spleen cell suspensions were prepared after sacrificing the mice, and cocultured with human peripheral blood lymphocytes that were prestimulated with PHA and IL-2 [104]. Supernatant samples were tested for the production of p24 antigen in ELISA after 1 and 2 weeks of culture. If no p24 antigen was detected (< 30 pg/ml) in the last sample, the tissue was designated as not infected. To detect integrated copies of the HIV genome, a PCR amplification of HIV *gag* sequences was performed on DNA isolated from spleen and peritoneal-wash cells [104]. The sensitivity of the PCR assay used is ~10 viral copies.

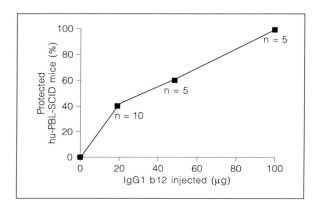

Fig. 5. Dose-response of protection against HIV-1 SF-2 infection by IgG1 b12. The percentage of protected hu-PBL-SCID mice is plotted against the dose of IgG1 b12 injected. The dose indicated was infected twice peritoneally at − 7 days and − 24 h relative to viral challenge. The number of mice used to calculate each point is indicated. After Parren et al. [104].

All animals injected with control antibodies Fab b13 or IgG1 HSV8 became infected with HIV-1. Two of eight mice injected with two doses of 20 µg neutralizing Fab b12 did not become infected. Passive transfer of antibody IgG1 b12 showed better protection with increasing doses: four of 10 mice at 20 µg, three of five at 50 µg and all five mice at 100 µg were found to be protected.

In figure 5, the protection by antibody IgG1 b12 is plotted as an in vivo protection dose-response curve. Increasing doses of IgG1 b12 show stronger protection, with complete protection in all mice at a 2×100 µg dose per mouse. Using this regimen and taking into account an antibody half-life of 7 days, we estimate the antibody dose at the time of virus challenge to be 4.5–7 mg/kg. The data from the pharmacokinetic study were used to calculate the serum IgG1 b12 concentration with this dose to be 135 ± 23 µg/ml (n = 5) at the time of virus injection. This value is ~15 × higher than the concentration needed to obtain 99% reduction of infectivity in vitro (< 10 µg/ml), as determined in p24 ELISA in a PBMC based assay. This indicates that antibody concentrations needed for complete protection against infection in this in vivo model, correspond with concentrations that neutralize virtually all virus as determined in in vitro assays, suggesting the protection experiments to be extremely stringent.

Whether and how the protection data can be translated to human HIV-1 infection is unclear at this time since there may be significant differences in the rates of viral replication, degree of T cell activation or availability of target cells between the hu-PBL-SCID model and infection in humans [132–134].

The calculated doses needed for protection however would be physiologically achievable doses in humans, but practically, may still be considered high. It is advantageous, however, that the half-life of human IgG1 in humans – generally between 21 and 23 days [135] – is three times longer than we found for human IgG1 in the mouse model. Furthermore, we have been able to increase the affinity of antibody b12 by mutagenesis and selection (see below) which improved both neutralization potency and breadth.

Antibodies to Other Epitopes on gp120

Epitope Masking
Previously, eight HIV-1 libraries were panned on gp120 coated directly to microtiter plates, resulting in Fabs specific for CD4bs-related epitopes. To extend the repertoire of human Fabs to a range of other epitopes, different selection strategies were employed as indicated in figure 6 [136].

Fabs to CD4bs/V2 Loop-Sensitive Epitope. We first adopted an epitope-masking strategy in which the CD4bs of gp120 was liganded with excess anti-CD4bs Fabs prior to panning (fig. 6) [137]. Following panning, three Fabs recognizing a novel epitope distinct from that recognized by conventional CD4bs antibodies were obtained. They were defined by the following criteria: (1) V2 region dependence indicated by sensitivity to amino acid changes in the V2 loop and by competition with murine monoclonal anti-V2 antibodies; (2) CD4bs dependence indicated by sensitivity to amino acid changes usually associated with CD4 binding and by inhibition of Fab binding to gp120 by soluble CD4. This dependence appeared to arise via conformational changes rather than direct binding since CD4bs antibodies (including b12) enhanced binding of two of the novel Fabs and, in a reversal of the competition format, the novel Fabs did not inhibit soluble CD4 binding to gp120; (3) equivalent binding to glycosylated and deglycosylated gp120 and significant, although much reduced binding, to denatured gp120 in contrast to CD4bs antibodies which do not bind to deglycosylated or denatured gp120. One of the novel Fabs efficiently neutralized the MN and IIIB strains of HIV-1 (1–10 µg/ml for 50% neutralization dependent on the assay used). These results indicate the presence of a novel neutralizing conformational epitope on gp120 sensitive to the V2 loop and the CD4bs. They further highlight the conformational flexibility of gp120 and emphasize that there is a relationship between the V2 loop and the CD4bs.

Fabs to the N-Terminus of gp120. The second strategy involved masking of the CD4-binding site by capturing gp120 on soluble CD4 or a murine anti-gp120 CD4-binding site mAb immobilized on microtiter wells [136]. Panning

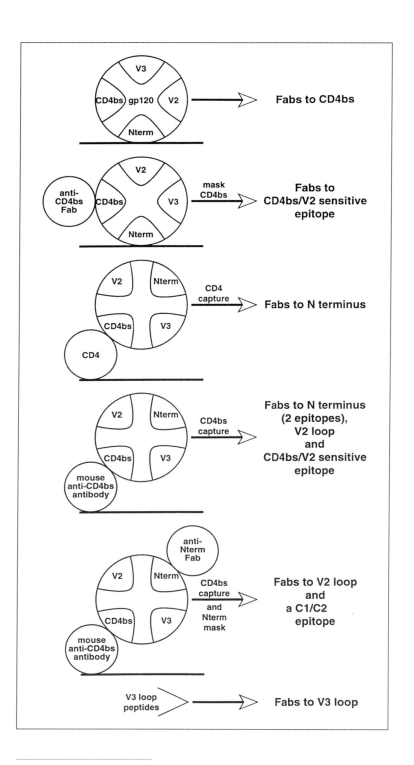

of the libraries on sCD4 captured gp120 resulted in the isolation of three novel Fab fragments (p7, p20, and p35). To map the epitope recognized, binding of the Fab fragments was assessed against a panel of HXBc2 gp120 mutants expressed in COS-1 cells [138, 139]. Binding of these Fabs is abolished by amino acid substitution 45 W/S in the C1 region of gp120, all three Fabs show also some dependency on substitutions at the C-terminus of gp120. Interestingly, these Fabs bind very poorly to gp120 directly coated to the plate. By panning on gp120 captured on anti-CD4bs mAb we isolated an additional set of Fab fragments (L19, L34, L35, L52, L59, L69, and L81). With the exception of L19 and L81, these Fabs recognize very similar epitopes as the three N-terminal Fabs above. Binding of L81, for example, is also impaired by mutations in the C3 and C5 region, in addition to the 45 W/S mutation. This indicates that at least two different epitope specificities can be distinguished within Fabs directed to the N-terminus of gp120. Binding of some of the Fabs is abolished by coating gp120 directly to the plate indicating a preferential orientation of gp120 upon coating, masking part of the N-terminal epitope.

V2 Loop of gp120. A second group of two Fabs was retrieved by panning against gp120 captured on anti-CD4bs mAb. These Fabs are specific to the V2 loop, indicated by sensitivity by substitutions in the V1/V2 loop and competition with a set of rodent anti-V2 loop antibodies [140].

C1/C2 Epitope. To demonstrate the ability to sequentially mask epitopes on an antigen to retrieve antibodies with new specificities, CD4bs mAb-captured gp120 was masked with a high-affinity anti-N terminal region Fab (Fab p7) prior to panning [136]. A Fab, L100, was retrieved which binds to a complex epitope involving the C1 and C2 regions of gp120. Substitutions 69 W/L and 76 P/Y abolish the binding of Fab L100, implying that the antibody binds to a part of the C1 region distinct from that recognized by the masking antibody p7. Substitutions 252 R/W, 256 S/Y, 262 N/T, and 267 E/L abolish or strongly impair the binding of Fab L100 implying the direct or indirect involvement of the C2 region in the epitope recognized.

Fig. 6. Strategies for panning phage display libraries against gp120. Most antibodies obtained by panning the libraries on recombinant gp120 directly coated to the surface of a microtiter plate are directed to CD4bs-related epitopes on gp120. To direct the panning to other epitopes we employed an epitope-masking technique. The CD4bs was blocked by addition of anti-CD4bs mAb or by capture of gp120 on soluble CD4 or anti-CD4bs mAb. Another set of antibodies was obtained by blocking with an Fab directed against N-terminal epitopes, or by panning against peptides. Antibody specificities obtained by use of these different panning strategies are indicated.

Selection of HIV-1 immune libraries against gp120 yields overwhelmingly antibodies reactive with the CD4bs. By refocussing this selection to other epitopes by capturing the gp120 ligand on its receptor CD4 or on anti-CD4bs mAb, a large number of antibodies were retrieved against the N-terminus of gp120. Such antibodies might have been missed in previous panning because gp120 seems to coat preferentially in a certain orientation occluding part of this epitope. The majority of these new antibodies show no or poor neutralizing ability even for T-cell-line-adapted viruses. Further, these antibodies bind very poorly to native multimer expressed on the cell surface of infected cells as measured in FACS analysis [131]. The affinities to monomeric gp120 as measured with BIAcore are high ($K_a > 10^8$), suggesting they do result from antigen-driven processes. We suggest that this antigen is viral debris generated during rapid viral turnover [141, 142]. The antibody response to native virus may in fact be very limited.

Panning on Peptides

V3 Loop of gp120. Libraries have been panned against (a) a cyclic peptide, N=CH-(CH2)3CO[SISGPGRAFYTG]NCH2CO-Cys-NH2 corresponding to the central most conserved part of the B clade V3 loop of gp120 coupled to BSA (b) a linear peptide corresponding to 24 residues of the MN V3 loop (RP142) coupled to ovalbumin (c) the same as b, but corresponding to the IIIB V3 loop (RP135) and (d) a linear peptide corresponding to a 'consensus' B clade V3 loop ('PND' peptide). Two neutralizing Fab fragments have been selected. The first anti-V3 loop Fab, loop 2, obtained by panning against the cyclic peptide corresponding to a 'consensus' V3 loop, has been converted to a whole IgG1 molecule. It is potent against MN and two primary isolates studied to date. The epitope recognized by loop 2 is probably GPGRAF. Loop 2 recognizes about 60% (10/17) of US clade B monomeric gp120 molecules. The second Fab, DO142-10, was selected by panning against the RP142 peptide and independently by panning against gp120 from the MN strain. Fab DO142-10 bound equally well to RP142 peptide and gp120 MN, but with lower affinity to the universal consensus peptide (UNI-V3), a JR-CSF V3 region fusion protein, and a recombinant gp120 from primary isolate W61D. No binding was observed with gp120 IIIB or a set of unrelated control antigens. Fab DO142-10 potently neutralizes T-cell-line-adapted strain HIV-1 MN, but not IIIB.

V1 Loop of gp120. The V1 loop may contain neutralizing epitopes as indicated from studies with sera of laboratory workers infected with HIV-1 IIIB [119], no mAbs to this region are available, however, which might be due to extreme variability of this region. Four linear 26-mer peptides from Dr. Seth Pincus (NIAID Rocky Mountains Laboratory) corresponding to the

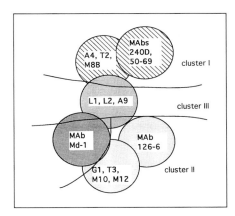

Fig. 7. Schematic summary of the mapping of antibodies binding to gp41. Circles group together antibodies with very similar competition profiles. Overlap between circles indicates competition between antibodies denoted within the respective circles. The different competition groups are conveniently organized into three clusters: cluster I (hatched circles) (residues 584–609); cluster II (light gray circles) (residues 649–668); and cluster III (dark gray circle). Md-1 is not grouped into any of the three clusters shown as it does not clearly belong to any of them, despite some overlap with clusters II and III. After Binley et al. [143].

sequences in gp120 IIIB and 3 primary isolates were used for panning. Serum titers to these peptides were weak and no positive clones were isolated from the human libraries or from a library prepared from bone marrow of a HIV-1 IIIB infected chimpanzee.

Antibodies to gp41

A large panel of human Fab fragments against the gp41 subunit of the HIV-1 envelope glycoprotein was isolated by panning six phage-displayed antibody libraries against recombinant gp41. Twenty-three Fabs recognizing conformation-dependent determinants on gp41 were isolated. Further selection of libraries against (1) gp41 ligated with Fabs from the initial selection and against (2) a recombinant gp41-containing gp140 protein yielded five additional Fabs. Competition of members of the Fab panel with one another and with previously described antibodies revealed a series of overlapping specificities which were conveniently grouped into three major epitope clusters (fig. 7). The majority of Fabs recognized epitopes involving residues 649–668 (previously designated the cluster II region), numbered using the Los Alamos IIIB/LAI sequence. A second set of Fabs reacted with an epitope involving residues 584–609 (known as the cluster I region). Another set of Fabs appeared to recognize a third and novel conformational epitope which has been termed the cluster III region. This third Fab epitope group demonstrated some overlap with both clusters I and II in binding assays. None of the Fabs neutralized HIV-1 laboratory strains at biologically significant concentrations. This tends to support the opinion that a vaccine based on the gp41 molecule has the drawback that neutralizing epitopes of gp41 are rare and/or unfavorably presented to the immune system [143].

Analysis of variable domain sequences revealed an interesting motif in the heavy chain CDR3s (HCDR3) of a set of six cluster III Fabs as follows: M/L (hydrophobic/aliphatic)-I-R-D/E (acidic/hydrophilic)-A/P (small/neutral/weakly hydrophobic)-L/F (hydrophobic)-D-L/F/V (hydrophobic). Another case of heavy chain CDR3 homology is present in the heavy chain sequences of 2 cluster II Fabs, which display the motif G-x-x-F-Y-S-G-x-G-R-R-x-x-F (where an x denotes a nonhomologous residue).

It is interesting that the two instances of HCDR3 homology occur in Fabs which are of very high affinity, especially the cluster II Fabs, which bind at approximately 4.5×10^9 M^{-1}. This affinity is approximately ten times that of gp120 Fabs that recognize a conformational epitope overlapping the CD4bs generated by similar techniques [144]. It is also notable that all eight Fabs which exhibit HCDR3 homologies to other Fabs have the same CDR1/CDR2 cononical structure [1–3], which is perhaps surprising considering these Fabs arise from a total of at least five separate antibody rearrangement events, and their HCDR3 homologies are probably generated during affinity maturation [145]. A common CDR1/CDR2 canonical structure implies a common structure in the heavy-chain binding region. To determine whether this observation means that such a main chain structure is predisposed to the 'convergent evolution' of functional antibody heavy-chain CDR3s or the CDR1/CDR2 is involved in the interaction with gp41 requires further study.

It has been demonstrated that an antibody frequently retains its specificity when the original heavy chain is paired with a number of different light chains, implying a dominant role for the heavy chain in the recognition of antigen in many instances [125]. Furthermore, it has been mentioned above that two of the three epitope clusters found are very similar to those described for antibodies cloned by classical methods, implying that the natural light chain partner is not essential for retention of the original epitope recognition. It has been documented by X-ray diffraction studies that the HCDR3 plays an important role in antibody specificity and this arises from its central location in the binding site [145]. These findings probably occur as a consequence of the conservation of antigenic elements of gp41 sequence and structure that are in turn maintained for a structural role in gp41-gp120 and gp41-gp41 interactions. Indeed, the majority of gp41 epitopes have been demonstrated to be obscured by gp120 by flow cytometry using infected cells and only become exposed upon shedding of the gp120 molecule which explains their lack of neutralizing efficacy. The gp41 regions recognized by the Fabs are likely inaccessible on the infectious virion and hence it is probable that these antibodies were generated not against intact virions, but against spikes where the gp120 subunit has been stripped, or precursor viral envelope protein or viral debris, which resulted in exposure of conserved gp41 structures [146]. It seems reasonable that the gp41 epitope structures are not subject to antigenic drift to escape the serum-neutralizing response, so allowing the antibody-antigen response to be 'honed to perfection' resulting in very high affinity antibodies to the most immunogenic regions, and a consequence of this has been the oligoclonal convergent evolution of HCDR3s in antibodies where the binding is predominantly mediated by the HCDR3. This interpretation rests on the understanding that these antibodies have been cloned from the natural antibody response to a very long-term persistent viral infection allowing for many generations of affinity maturation.

Convergent evolution resulting in a common HCDR3 motif sequences in several antibodies, therefore may be an interesting consequence of a persistent immune response to conserved antigen structures.

In vitro Evolution of Fabs to Improve Affinity and Strain Cross-Reactivity and Increase Neutralization Potency

A strategy has been developed for the improvement of antibody affinity in vitro [147]. The method termed CDR walking, does not require structural information on either antibody or antigen. CDRs are targeted for random mutagenesis followed by selection for increased affinity for antigen by panning of the appropriate phage display library.

We have applied the method to potent neutralizing CD4bs anti-gp120 Fab b12. In the first experiment, the entire CDR1 of the heavy chain (HCDR1) was targeted for mutagenesis using an overlap PCR mutagenesis strategy. Briefly, a random oligonucleotide sequence (NNS where N = A, T, G or C and S = G or C) was incorporated by overlap PCR to replace the existing HCDR1 and generate a phage display library of Fab b12 with essentially all amino acid residue combinations present in HCDR1. This library was panned for four rounds against gp120 (IIIB). The surviving clones were then mutagenized at positions 96–99 of HCDR3; a region suggested to be a 'hot spot' by chain-shuffling experiments [125]. The new library was then panned for six rounds against gp120 IIIB, after which a strong consensus was observed in the sequences of both CDRs from surviving clones. Further, the affinities of the Fabs were improved for binding to both IIIB and MN gp120. The best Fab (Fab 3B3) had an 8-fold-enhanced affinity for IIIB gp120 and a 6-fold-enhanced affinity for MN gp120 as measured by surface plasmon resonance. The enhanced affinity was reflected in enhanced neutralization potency of approximately an order of magnitude towards the MN strain of HIV-1. A

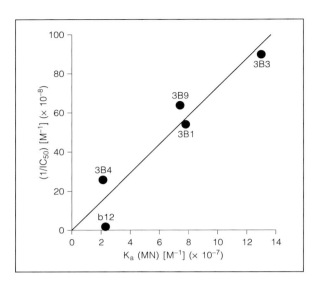

Fig. 8. Correlation of affinity increase with neutralization potency against HIV-1 MN of a set of Fabs evolved from Fab b12. The reciprocal IC_{50} value obtained with HIV-1 MN in a quantitative infectivity assay based on syncytium formation is plotted against the K_a for binding gp120 MN as determined by BIAcore. The highest affinity Fab 3B3 is 6-fold improved in affinity and 54-fold improved in neutralization potency as compared with parent Fab b12. After Barbas et al. [147].

linear correlation was found between increase of affinity and neutralization potency (fig. 8). Furthermore the most improved Fab (3B3) was shown capable of neutralizing a number of primary clinical isolates of HIV-1 which were not neutralized by the parent Fab (b12).

Therefore, by targeting a very small number of residues in two CDRs, we have been able to improve neutralization potency by roughly an order of magnitude. The affinity of Fab b12 was further improved for its affinity to gp120 by preparing five additional libraries by saturation-mutagenesis of CDRs and selection on immobilized gp120 [148]. Sequential and parallel optimization strategies of CDRs were examined. The sequential CDR walking strategy consistently yielded b12 variants of improved affinity in each of the four different optimization sequences examined, resulting in a 96-fold improvement of affinity. Additivity effects in the antibody combining site were explored by combining independently optimized CDRs in a parallel optimization strategy. Six variants containing optimized CDRs were constructed. Improvement of affinity based on additivity effects proved to be unpredictable but did lead to a modest improvement in affinity in only one of the six combinations. The highest affinity Fab prepared using this strategy was improved 420-fold in affinity. The affinity of this Fab was 15 pM as compared to 6.3 nM for b12. Examination of the kinetics of Fab binding to gp120 revealed that improvements in affinity were dominated by a slowing of the off-rate of the Fab.

Fabs with up to 420-fold-improved affinities would have obvious advantages for applications in the prophylaxis or immunotherapy of HIV-1 infection, if the correlation between affinity and neutralization potency against this epitope remained linear as shown for the Fab in the first in vitro evolution experiment. The effective in vivo protective dose of such antibodies would be much smaller. The relationship between neutralization potency and affinity for these very-high-affinity Fab fragments, however, appears to be more complex than could be expected from the previous studies and is still under investigation. Finally, it should be noted that the observed correlation between affinity to monomeric gp120 and neutralization potency for b12 may pertain only to this antibody. Antibody b12 recognizes a potent neutralizing epitope associated with binding to native HIV-1 envelope oligomer (see below), and the increased affinity to the gp120 monomer of improved Fab 3B3 correlates with increased affinity for HIV-1 envelope oligomer expressed on the surface of infected cells [149]. Studies with other antibodies against epitopes overlapping the CD4bs (below), or other gp120 epitopes indicated however that binding to recombinant gp120 monomer generally does not correlate with neutralization of HIV-1 [78, 144].

Recognition Properties of Fab b12 Indicate Efficient Binding to Native HIV-1 Envelope Oligomer

The potent HIV-1 neutralizing ability of antibody b12 and its broad cross-clade reactivity suggest the presence of a conserved structurally important feature on infectious virions associated with the CD4bs. Identification of this feature has important implications for immunotherapy and vaccine design.

Comparing b12 with a panel of 5 other anti-CD4bs Fabs, showing a spectrum of neutralizing abilities, indicated that potent neutralization does not generally correlate with affinity for recombinant gp120, or the ability of an Fab to compete with soluble CD4. Epitope mapping by cross-competition using ELISA or surface plasmon resonance suggests that all six Fabs bind to an overlapping site [144]. The recognition of a panel of HIV-1 gp120 mutants by the six Fabs showed that patterns of sensitivity to particular gp120 amino acid changes were similar for all six Fabs to those seen for anti-CD4bs mAbs derived from HIV-1-infected individuals by conventional means. In addition, recognition by Fab b12 demonstrated an atypical sensitivity to changes in the V1/V2 variable regions. The difference between Fab b12 and the other Fabs

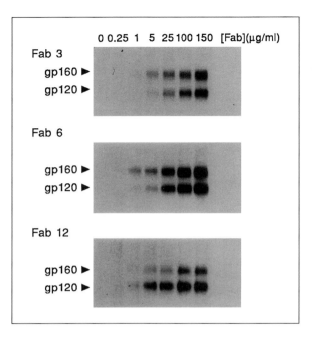

Fig. 9. Fab binding to envelope glycoproteins expressed on the surface of COS-1 cells. COS cells were transfected with a plasmid expressing the envelope glycoprotein of the HXBc2 molecular clone of HIV-1 IIIB. The cells were [^{35}S]-cysteine radiolabeled for 2 days after transfection and incubated with the Fabs indicated. After three washes with PBS, the cells were lysed in Nonidet P40 buffer and Fab-envelope glycoprotein complexes were precipitated with goat anti-human F(ab')$_2$ and protein G-Sepharose and analyzed by SDS-PAGE. Autoradiographs of immunoprecipitated complexes of Fabs b3, b6, and b12 are shown. After Roben et al. [144].

most likely to be directly related to neutralization potency was found when binding to cell surface envelope was assessed in radio-immunoprecipitation studies. As shown in figure 9, b12 appears to bind more efficiently to cell surface gp120 at low concentrations than the other less potently neutralizing Fabs, b3 and b6. Furthermore, b12 appears to bind preferentially to cell surface mature gp120 whereas b3 and b6 bind preferentially to gp160. We suggest that mature gp120, which is present on the cell surface as an oligomer, best approximates the functionally relevant envelope glycoprotein complex and preferential binding by b12 leads to more efficient neutralization of virus. The CD4bs on monomeric gp120, therefore, presumably provides an array of overlapping epitopes, many of which are less accessible to antibody on the oligomeric membrane associated form of gp120.

Table 2. Binding of recombinant Fabs to HIV-1 MN-infected cells and HIV-1 neutralization

Fab fragment	Concentration at half-maximal fluorescence µg/ml	HIV-1-neutralizing ability
Fab b12	0.07	potent
Fab b11	0.7	intermediate
Fab b14	2.0	weak
Fab b3	4.5	weak

We approached this also in another way, by titrating a panel of Fab with differing neutralizing ability on cells freshly infected with HIV-1 MN and analysis on FACS [131]. Table 2 shows the concentrations of Fabs at half maximal fluorescence intensity. It is clearly demonstrated that Fab b12 binds with much higher affinity (approximately 100-fold) than Fabs which display lesser neutralizing abilities. Again it should be stressed that no difference in affinity was found between these Fab fragments when binding to recombinant monomeric gp120 was compared.

We postulate that Fab b12 binds very efficiently to an epitope available on the surface of the native HIV-1 envelope glycoprotein, which leads to effective neutralization of the virus.

Antibody Phage Display Libraries as Tools to Assess Vaccines

Potential viral vaccines are often evaluated by immunization of animals and measurement of the functional activity of serum antibodies in vitro. This is followed by animal protection studies and finally trials in humans. We propose that antibody phage display libraries could be used as an additional approach to obtain a rapid assessment of the exposure of neutralizing epitopes on the candidate vaccine in vitro. Antibody libraries prepared from immunized donors, containing a mixture of neutralizing and non-neutralizing antibodies displayed on the surface of phage, are challenged with the vaccine preparation. We propose that on a good vaccine neutralizing epitopes should be expressed most favorably, leading to specific selection of neutralizing antibodies in a panning experiment.

We applied this approach to compare three preparations of recombinant HIV-1 envelope; monomeric gp120, gp160 and a soluble gp140 oligomer. The gp140 oligomer was prepared by removing the gp120/gp41 cleavage site and

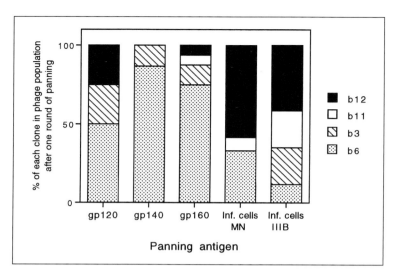

Fig. 10. A mini-phage display library was constructed containing equal amounts of four different anti-CD4bs phage-Fabs (b6, b3, b11, and b12). The library was panned for one round against the five HIV-1 envelope glycoprotein preparations indicated, and the representation of the clones was determined. Approximately twenty clones were screened for the data shown in each bar. After Parren et al. [152].

gp41 transmembrane domain [150, 151], and may therefore fold in a configuration different from native membrane-associated oligomer. None of the preparations efficiently selected neutralizing antibodies from antibody phage display libraries prepared from long-term asymptomatic HIV-1 seropositive donors [152] although the libraries do contain such antibodies directed to the CD4bs of gp120, the V3 loop and a CD4bs/V2 loop-sensitive epitope [124, 125, 127, 137]. We specifically investigated the gp120 CD4bs epitope in greater detail with a control library using a mixture of four phage-displaying well-character-ized anti-CD4bs Fabs (fig. 10). Recombinant gp160 and gp140 multimer preferentially selected weakly neutralizing Fabs from this minilibrary. Gp120 selected a higher proportion of a potent neutralizing Fab. The most favorable result was found however with native oligomeric HIV-1 envelope, as expressed on infected cells, which selected the potent neutralizing Fab preferentially.

What advantages does the library approach offer in evaluating a vaccine compared to, for example, simply measuring the ability of a panel of existing antibodies to bind the vaccine in question? First, one doesn't need to have any a priori knowledge of the antibodies selected: after several rounds of panning, a polyclonal preparation of Fabs may be prepared to test for the neutralizing ability of the mixture of individual Fabs selected. In this study,

Table 3. Anti-HIV-1 envelope glycoprotein specificities
obtained from phage display libraries

	Selecting antigen
Anti-gp120	
CD4bs	rgp120, rgp160, rgp 140 tetramer
V3 loop	peptides, rgp120
V2 loop	rgp120 with epitope masking
V2 loop/CD4bs sensitive	rgp120 with epitope masking
C1 (2 epitopes)	rgp120 with epitope masking
C1/C2	rgp120 with epitope masking
Anti-gp41	
Cluster I	rgp41, rgp140 tetramer
Cluster II	rgp41, rgp140 tetramer
Cluster III	rgp41

we deliberately chose a well-characterized system as model. Normally, such detailed information on the antibodies would not be so readily available. Second, the antigen preparation may select for totally new antibodies against previously uncharacterized epitopes. The approach then leads directly to mAbs of great interest. Third, during the panning procedure, antibodies present in the library may bind simultaneously to distinct epitopes as in an 'in vivo' situation, in which, for example, synergies and induced conformational changes may play a role. Fourth, information is obtained directly on human antibodies reacting with the vaccine: murine antibodies commonly available may reflect peculiarities of immunization in the animal model.

In summary, the presence of antibodies in a phage display library against overlapping epitopes with varying neutralizing ability mimics the in vivo situation. The immune repertoire can then be challenged in vitro by selecting the library against a potential vaccine. For a good vaccine, neutralizing epitopes should be most favorably presented leading to selection of neutralizing antibody clones.

Conclusions

Antibodies from phage display libraries have been used to map humoral immune responses to gp120 and gp41 in HIV-1 infection. These antibody responses seem to be most relevant since all HIV-1-neutralizing epitopes identified to date map on these two antigens. We have identified antibodies against most major epitopes identified by mAb obtained with conventional methods, and we identified several novel epitopes. Table 3 lists all the specificities obtained.

Interestingly, the majority of the antibodies obtained are directed against non- or weakly-neutralizing epitopes, and bind poorly to HIV-1 envelope glycoprotein as expressed on the surface of infected cells. These antibodies, however, all bind with high affinity to recombinant monomeric gp120 as measured by surface plasmon resonance, which strongly suggest that they were generated by an antigen-driven process. Most of the antibody response may therefore be generated against viral debris, while the response to native virus seems very limited.

IgG1 b12, obtained from phage display libraries as an Fab and converted into a whole antibody, is one out of only three antibodies which are currently considered to have clinical potential. IgG1 b12, further, is the only potent neutralizing and broadly cross-clade-reactive antibody directed against the CD4bs available. The two other antibodies are anti-gp120 mAb 2G12 and anti-gp41 mAb 2F5 [118, 120, 121, 130, 153]. The three mAbs are directed against distinct epitopes and do not compete and could therefore be combined in a cocktail for use in immunoprophylaxis and immunotherapy to decrease viral load in infected individuals. Such possibilities are currently being explored.

Acknowledgments

We are grateful to many valued collaborators and to members of our Laboratory for their contributions to this work, including Carlos Barbas, James Binley, Henrik Ditzel, Paola Fisicaro, Meng Wang, and Anthony Williamson. Some of the work described is supported by the National Institutes of Health (AI33292). P.W.P. is the recipient of a scholarship award from the Pediatric AIDS Foundation (No. 77290–20–PF).

References

1   Behring E, Kitasato S: Ueber das Zustandekommen der Diphterie-Immunität und der Tetanus-Immunität bei Thieren. Dt Med Wochenschr 1890;16:1113.
2   Behring EA: Die Geschichte der Diphtherie. Leipzig, Thieme, 1983, p 186.
3   Behring EA: Das neue Diphtherieheilmittel. Berlin, Hering, 1894, p 40.
4   Bowman JM: The prevention of Rh immunization. Trans Med Rev 1983;2:129–150.
5   Enria DA, Briggiler AM, Fernandez NJ, Levis SC, Maiztegui JI: Importance of dose of neutralising antibodies in treatment of Argentine haemorrhagic fever with immune plasma. Lancet 1984;ii: 255–256.
6   Kohler G, Milstein C: Continuous cultures of fused cells secreting antibody of predefined specificity. Nature 1975;256:495–497.
7   Carson DA, Freimark BD: Human lymphocyte hybridomas and monoclonal antibodies. Adv Immunol 1986;38:275–311.

8   Casali P, Notkins AL: Probing the human B-cell repertoire with EBV: Polyreactive. Annu Rev Immunol 1989;7:513–535.
9   Parren PWHI: Preparation of genetically engineered monoclonal antibodies for human immunotherapy. Hum Antibody Hybridomas 1992;3:137–145.
10   Lum LG, Burns E, Janson MM: IgG anti-tetanus toxoid antibody synthesis by human bone marrow. I. Two distinct populations of marrow B cells and functional differences between marrow and peripheral blood B cells. J Clin Immunol 1990;10:255–264.
11   Boulianne GL, Hozumi N, Shulman MJ: Production of functional chimaeric mouse/human antibody. Nature 1984;312:643–646.
12   Morrison SL, Johnson MJ, Herzenberg LA: Chimeric human antibody molecules: Mouse antigen-binding domains with human constant domains. Proc Natl Acad Sci USA 1984;81:6851–6855.
13   Neuberger MS, Williams GT, Mitchell EB: A hapten-specific chimeric immunoglobulin E antibody which exhibits human physiological effector function. Nature 1985;314:268–271.
14   Bruggemann M, Williams GT, Bindon MR, Clark MR, Walker MR, Jefferis R, Waldmann H, Neuberger MS: Comparison of the effector functions of human immunoglobulins using a matched set of chimeric antibodies. J Exp Med 1987;166:1351–1361.
15   Parren PWHI, Geerts MEJ, Boeije LCM, Aarden LA: Induction of T-cell proliferation by recombinant mouse and chimeric mouse/human anti-CD3 monoclonal antibodies. Res Immunol 1991;142:749–763.
16   Jones PT, Dear PH, Foot J: Replacing the complementarity-determining regions in a human antibody with those from a mouse. Nature 1986;321:522–525.
17   Verhoeyen M, Milstein C, Winter G: Reshaping human antibodies: Grafting an antilysosyme activity. Science 1988;239:1534–1536.
18   Co MS, Queen C: Humanized antibodies for therapy. Nature 1991;351:501–502.
19   Bruggemann M, Winter G, Waldmann H, Neuberger MS: The immunogenicity of chimeric antibodies. J Exp Med 1989;170:2153–2157.
20   Riechmann L, Clark M, Waldmann H, Winter G: Reshaping human antibodies for therapy. Nature 1988;332:323–327.
21   Better M, Chang CP, Robinson RR, Horwitz AH: *Escherichia coli* secretion of an active chimeric antibody fragment. Science 1988;240:1041–1043.
22   Orlandi R, Gussow DH, Jones PT, Winter G: Cloning immunoglobulin variable domains for expression by the polymerase chain reaction. Proc Natl Acad Sci USA 1989;86:3833–3837.
23   Larrick JW, Danielsson L, Brenner CA, Abrahamson M, Fry KE, Borrebaeck CAK: Rapid cloning of rearranged immunoglobulin genes from human hybridoma cells using mixed primers and the polymerase chain reaction. Biochem Biophys Res Commun 1989;160:1250–1256.
24   Huse WD, Sastry L, Iverson SA, Kang AS, Alting-Mees M, Burton DR, Benkovic SJ, Lerner RA: Generation of a large combinatorial library of the immunoglobulin repertoire in phage lambda. Science 1989;246:1275–1281.
25   Persson MA, Caothien RH, Burton DR: Generation of diverse high-affinity human monoclonal antibodies by repertoire cloning. Proc Natl Acad Sci USA 1991;88:2432–2436.
26   Mullinax RL, Gross EA, Amberg JF, Hay BN, Hogrefe HH, Kubitz MM, Greener A, Alting-Mees M, Ardourel D, Short JM, Sorge JA, Shopes B: Identification of human antibody fragment clones specific for tetanus toxoid in a bacteriophage lambda immunoexpression library. Proc Natl Acad Sci USA 1990;87:8095–8099.
27   Burton DR, Barbas III CF: Human antibodies from combinatorial libraries. Adv Immunol 1994; 57:191–280.
28   Smith GP: Filamentous fusion phage: Novel expression vectors that display cloned antigens on the virion surface. Science 1985;228:1315–1317.
29   Scott JK, Smith GP: Searching for peptide ligands with an epitope library. Science 1990;249:386–390.
30   Devlin JJ, Panganiban LC, Devlin PE: Random peptide libraries: A source of specific protein binding molecules. Science 1990;249:404–406.
31   Cwirla SE, Peters EA, Barrett RW, Dower WJ: Peptides on phage: A vast library of peptides for identifying ligands. Proc Natl Acad Sci USA 1990;87:6378–6382.
32   McCafferty J, Griffiths AD, Winter G, Chiswell DJ: Phage antibodies: Filamentous phage displaying antibody variable domains. Nature 1990;348:552–554.

33 Kang AS, Barbas III CF, Janda KD, Benkovic SJ, Lerner RA: Linkage of recognition and replication functions by assembling combinatorial antibody Fab libraries along phage surfaces. Proc Natl Acad Sci USA 1991;88:4363–4366.

34 Barbas III CF, Kang AS, Lerner RA, Benkovic SJ: Assembly of combinatorial antibody libraries on phage surfaces: The gene III site. Proc Natl Acad USA 1991;88:7978–7982.

35 Stevens RH, Macy E, Morrow C, Saxon A: Characterization of a circulating subpopulation of spontaneous antitetanus toxoid antibody producing B cells following in vivo booster immunization. J Immunol 1979;122:2498–1504.

36 Thiele CJ, Morrow CD, Stevens RH: Multiple subsets of anti-tetanus toxoid antibody-producing cells in human peripheral blood differ by size, expression of membrane receptors, and mitogen reactivity. J Immunol 1981;126:1146–1153.

37 Ershler WB, Moore AL, Hacker MP: Specific in vivo and in vitro antibody response to tetanus toxoid immunization. Clin Exp Immunol 1982;49:552–558.

38 Volkman DJ, Allyn SP, Fauci AS: Antigen-induced in vitro antibody production in humans: Tetanus toxoid specific antibody synthesis. J Immunol 1982;129:107–112.

39 Barbas SM, Ditzel HJ, Salonen EM, Wei-Ping Y, Silverman GJ, Burton DR: Human autoantibody recognition of DNA. Proc Natl Acad Sci USA 1995;92:2529–2533.

40 Willaimson RA, Burioni R, Sanna PP, Partridge LJ, Barbas III CF, Burton DR: Human monoclonal antibodies against a plethora of viral pathogens from single combinatorial libraries. Proc Natl Acad Sci USA 1993;90:4141–4145.

41 Barbas III CF: Recent advances in phage display. Curr Opin Biotechnol 1993;4:526–530.

42 Barbas SM, Barbas III CF: Filamentous phage display. Fibrinolysis 1994;8(suppl 1):245–252.

43 Cortese R, Monaci P, Nicosia A, Luzzago A, Felici F, Galfré G, Pessi A, Tramontano A, Sollazzo M: Identification of biologically active peptides using random libraries displayed on phage. Curr Opin Biotechnol 1995;6:73–80.

44 Cortese R, Felici F, Galfré G, Luzzago A, Monaci P, Nicosia A: Epitope discovery using peptide libraries displayed on phage. Trends Biotechnol 1994;12:262–266.

45 Fong S, Doyle LV, Devlin JJ, Doyle MV: Scanning whole cells with phage-display libraries: Identification of peptide ligands that modulate cell function. Drug Dev Res 1994;33:64–70.

46 Gallop MA, Barrett RW, Dower WJ, Fodor SPA, Gordon EM: Applications of combinatorial technologies to drug discovery. 1. Background and peptide combinatorial libraries. J Med Chem 1994;37:1233–1251.

47 Gordon EMM, Barrett RW, Dower WJ, Fodor SPA, Gallop MA: Applications of combinatorial technologies to drug discovery. 2. Combinatorial organic synthesis, library screening strategies, and future directions. J Med Chem 1994;37:1385–1401.

48 Scott JK, Craig L: Random peptide libraries. Curr Opin Biotechnol 1994;9:40–48.

49 Winter G, Griffiths AD, Hawkins RE, Hoogenboom HR: Making antibodies by phage display technology. Annu Rev Immunol 1994;12:433–455.

50 Walker NR, Lund J, Thompson KM, Jefferis R: Aglycosylation of human IgG1 and IgG3 monoclonal antibodies can eliminate recognition by human cells expressing FcgammaRI and/or FcgammaRII receptors. Biochem J 1989;259:347–353.

51 Tao MH, Morrison SL: Studies of aglycosylated chimeric mouse-human IgG: Role of carbohydrate in the structure and effector functions mediated by the human IgG constant region. J Immunol 1989;143:2595–2601.

52 Wright A, Shin SU: Production of genetically engineered antibodies in myeloma cells: Design, expression and applications; in Lerner A, Burton DR (eds): Methods: A Companion to Methods in Enzymology. Orlando, Academic Press, 1991, pp 125–135.

53 Bebbington CR: Expression of antibody genes in nonlymphoid mammalian cells; in Lerner RA, Burton DR (eds): Methods: A Companion to Methods in Enzymology. Orlando, Academic Press, 1991.

54 Hasemann CA, Capra JD: High-level production of a functional immunoglobulin heterodimer in a baculovirus expression system. Proc Natl Acad Sci USA 1990;87:3942–3946.

55 Bender E, Woof JM, Atkin JD, Barker MD, Bebbington CR, Burton DR: Recombinant human antibodies: Linkage of an Fab fragment from a combinatorial library to an Fc fragment for expression in mammalian cell culture. Hum Antibody Hybridomas 1993;4:74–79.

56   Burton DR, Pyati J, Koduri R, Sharp SJ, Thornton GB, Parren PWHI, Sawyer LSW, Hendry RM, Dunlop N, Nara PL, Lamacchia M, Garratty E, Stiehm ER, Bryson YJ, Cao Y, Moore JP, Ho DD, Barbas III CF: Efficient neutralization of primary isolates of HIV-1 by a recombinant human monoclonal antibody. Science 1994;266:1024–1027.

57   Bebbington CR, Renner G, Thomson S, King D, Abrams D, Yarranton GT: High-level expression of a recombinant antibody from myeloma cells using a glutamine synthetase gene as an amplifiable selectable marker. Bio/Technology 1992;10:169–175.

58   Chanock RM, Crowe Jr JE, Murphy BR, Burton DR: Human monoclonal antibody Fab fragments cloned from combinatorial libraries: Potential usefulness in prevention and/or treatment of major human viral diseases. Infect Agents Dis 1993;2:118–131.

59   Krugman SR, Ward R, Giles JP, Jacobs AM: Infectious hepatitis: Studies on the effect of gamma globulin and on the incidence of inapparent infection. JAMA 1960;174:83–90.

60   Snydman DR: Cytomegalovirus immunoglobulins in the prevention and treatment of cytomegalovirus disease. Rev Infect Dis 1990;12:S839–S848.

61   Snydman DR, Werner BG, Heinze-Lacy B, Beredi BP, Tilney NL, Kirkman RL, Milford EL, Cho SI, Bush HL, Levey AS, Strom TB, Carpenter CB, Levey RH, Harman WE, Zimmerman ZE, Shapiro ME, Steinman T, LoGerfo F, Idelson B, Schroter GPJ, Levin MJ, McIver J, Leszczynski J, Grady GF: Use of cytomegalovirus immune globulin to prevent cytomegalovirus disease in renal-transplant recipients. N Engl J Med 1987;317:1049–1054.

62   Groothuis JR, Simoes EAF, Levin MJ, Hall CB, Rodriguez WJ, Meissner C, Welliver RC, Hemming VG, RSVIG Study Group: Respiratory syncytial virus (RSV) immune globulin (IG) prevents severe lower respiratory infection (LRI) in high risk children. Pediatr Res 1993;33:169A.

63   Beasley RP, Hwang LY, Stevens CE, Lin CC, Hsieh FJ, Wang KY, Sun TS, Szmuness W: Efficacy of hepatitis B immune globulin for prevention of perinatal. Hepatology 1983;3:135–141.

64   Orenstein WA, Heymann DL, Ellis RL, Rosenberg RL, Nakano J, Halsey NA, Overturf GD, Hayden DF, Witte JJ: Prophylaxis of varicella in high risk children: Dose response effect of zoster immune globulin. J Pediatr 1981;98:368–373.

65   Loofbourow JC, Cabasso VJ, Roby RE, Anuskiewicz W: Rabies immune globulin (human) clinical trials and dose determination. JAMA 1971;217:1825–1831.

66   Berman PW, Gregory TJ, Riddle L, Nakamura GR, Champe MA, Porter JP, Wurm FM, Hershberg RD, Cobb EK, Eichberg JW: Protection of chimpanzees from infection by HIV-1 after vaccination with recombinant glycoprotein gp120 but not gp160. Nature 1990;345:622–625.

67   Girard M, Kieny MP, Pinter A, Barre-Sinousi F, Nara PL, Kolbe H, Kusumi K, Chaput A, Reinhart T, Muchmore E, Ronco J, Kaczorek M, Gomard E, Gluckman J-C, Fultz PN: Immunization of chimpanzees confers protection against challenge with human immunodeficiency virus. Proc Natl Acad Sci USA 1991;88:542–546.

68   Berman PW, Eastman DJ, Wilkes DM, Nakamura GR, Gregory TJ, Schwartz D, Gorse G, Belshe R, Clements ML, Byrn RA: Comparison of the immune response to recombinant gp120 in humans and chimpanzees. AIDS 1994;8:591–601.

69   Bruck C, Thiriart C, Fabry L, Francotte M, Pala P, Van Opstal O, Culp J, Rosenberg M, DeWilde M, Heidt P, Heeney J: HIV-1 envelope-elicited neutralizing antibody titres correlate with protection and virus load in chimpanzees. Vaccines 1994;12:1141–1148.

70   Berman PW, Eastman DJ, Nakamura GR, Obijeski JF, Francis D, Gregory TJ, Murthy KK: Apparent protection of MN-rgp120-immunized chimpanzees from infection with a primary isolate of HIV-1; in Chanock RM, Brown F, Ginsberg HS, Norrby E (eds): Vaccines 95; Molecular Approaches to the Control of Infectious Diseases. Plainview, Cold Spring Harbor Laboratory Press, 1995, pp 143–148.

71   Emini EA, Schleif WA, Nunberg JH, Conley AJ, Eda Y, Tokiyoshi S, Putney SD, Matsushita S, Cobb KE, Jett CM, Eichberg JW, Murthy KK: Prevention of HIV-1 infection in chimpanzees by gp120 V3 domain-specific monoclonal antibody. Nature 1992;355:728–730.

72   Giavedoni LD, Planelles V, Haigwood NL, Ahmad S, Kluge JD, Mathas ML, Gardner MB, Luciw PA, Yilma TD: Immune response of rhesus macaques to recombinant simian immunodeficiency virus gp130 does not protect from challenge infection. J Virol 1993;67:577–583.

73   Israel ZR, Edmondson PF, Maul DH, O'Neil SP, Mossman SP, Thiriart C, Fabry L, Van Opstal O, Bruck C, Bex F, Burny A, Fultz PN, Mullins JI, Hoover EA: Incomplete protection, but suppression of virus burden, elicited by subunit simian immunodeficiency virus vaccines. J Virol 1994; 68:1843–1853.

74   Siegel F, Kurth R, Norley S: Neither whole inactivated virus immunogen nor passive immunoglobulin transfer protects against SIVagm infection in the African green monkey natural host. J Acquir Immune Defic Syndr 1995;8:217–226.

75   Putkonen P, Thorstensson R, Ghavamzadeh L, Albert J, Hild K, Biberfeld G, Norrby E: Prevention of HIV-2 and SIV$_{sm}$ infection by passive immunization in cynomolgus monkeys. Nature 1991;352: 436–438.

76   Kent KA, Kitchin P, Mills KHG, Page M, Taffs F, Corcoran T, Silvera P, Flanagan B, Powell C, Rose J, Ling C, Aubertin AM, Scott EJ: Passive immunization of cynomolgus macaques with immune sera or a pool of neutralizing monoclonal antibodies failed to protect against challenge with SIVmac251. AIDS Res Hum Retroviruses 1994;10:189–194.

77   Levy J, Youvan T, Lee ML, Passive Hyperimmune Therapy Study Group: Passive hyperimmune plasma therapy in the treatment of acquired immunodeficiency syndrome: Results of a 12-month multicenter double-blind controlled trial. Blood 1994;84:2130–2135.

78   Moore JP, Cao Y, Qing L, Sattentau QJ, Pyati J, Koduri R, Robinson J, Barbas CF III, Burton DR, Ho DD: Primary isolates of human immunodeficiency virus type 1 are relatively resistant to neutralization by monoclonal antibodies to gp120, and their neutralization is not predicted by studies with monomeric gp120. J Virol 1995;69:101–109.

79   Crothers DM, Metzger H: The influence of polyvalency on the binding properties of antibodies. Immunochemistry 1972;9:341–357.

80   Hornick CL, Karush F: Antibody affinity – III the role of multivalence. Immunochemistry 1972; 9:325–340.

81   Mason DW, Williams AF: Kinetics of antibody reactions and the analysis of cell surface antigens; in Weir DM (ed): Handbook of Experimental Immunology. Oxford, Blackwell, 1986.

82   Pack P, Plückthun A: Miniantibodies: Use of amphipathic helices to produce functional, flexibly linked dimeric F$_V$ fragments with high avidity in *Escherichia coli*. Biochemistry 1992;31:1579–1584.

83   Keller R: The stability of neutralization of poliovirus by native antibody and enzymatically derived fragments. J Immunol 1966;96:96–106.

84   Icenogle J, Shiwen H, Duke G, Gilbert S, Rueckert R, Anderegg J: Neutralization of poliovirus by a monoclonal antibody: Kinetics and stoichiometry. Virology 1983;127:412–425.

85   Thomas AAM, Brioen P, Boeyé A: A monoclonal antibody that neutralizes poliovirus by cross-linking virions. J Virol 1985;54:7–13.

86   Ruggeri FM, Greenberg HB: Antibodies to the trypsin cleavage peptide VP8 neutralize rotavirus by inhibiting binding of virions to target cells in culture. J Virol 1991;65:2211–2219.

87   Yoden S, Kida H, Yanagawa R: Is bivalent binding of monoclonal antibodies to different antigenic areas on the hemagglutinin of influenza virus required for neutralization of viral infectivity? Virology 1985;85:209–216.

88   Dimmock NJ: Neutralization of animal viruses. Curr Top Microbiol Immunol 1993;183:1–149.

89   Buchmeier MJ, Lewicki HA, Talbot PJ, Knobler RL: Murine hepatitis virus-4 (strain JHM)-induced neurologic disease is modulated in vivo by monoclonal antibody. Virology 1984;132:261–270.

90   Talbot PJ, Salmi AA, Knobler PJ, Buchmeier MJ: Topographical mapping of murine hepatitis virus-4 (strain JHM): Correlation with biological activities. Virology 1984;132:250–260.

91   Rector JT, Lausch RL, Oakes JE: Use of monoclonal antibodies for analysis of antibody-dependent immunity to ocular herpes simplex virus type 12 infection. Infect Immun 1982;38:168–174.

92   Kümel G, Kaerner HC, Levine M, Schröder CH, Glorioso JC: Passive immune protection by herpes simplex virus-specific monoclonal antibodies and monoclonal antibody-resistant mutants altered in pathogenicity. J Virol 1985;56:930–937.

93   Balachandran N, Bacchetti S, Rawls WE: Protection against lethal challenge of balb/c mice by passive transfer of monoclonal antibodies to five glycoproteins of herpes simplex type 2. Infect Immun 1982;37:1131–1137.

94   Deregt D, Gifford GA, Ljaz MK, Watts TC, Gilchrist JE, Haines DM, Babiuk LA: Monoclonal antibodies to bovine coronavirus Glycoproteins E2 and E3: Demonstration of in vivo virus-neutralizing activity. J Gen Virol 1989;70:993–998.

95   Mathews JH, Roehrig JT, Trent DW: Role of complement and the Fc portion of immunoglobulin G in immunity to Venezuelan equine encephalomyelitis virus infection with glycoprotein-specific monoclonal antibodies. J Virol 1985;55:594–600.

96   Baldridge JR, Buchmeier MJ: Mechanisms of antibody-mediated protection against lymphocytic choriomeningitis virus infection: Mother-to-baby transfer of humoral protection. J Virol 1992;66: 4252–4257.

97   Schlesinger JJ, Foltzer M, Chapman S: The Fc portion of antibody to yellow fever virus NS1 is a determinant of protection against YF encephalitis in mice. Virology 1993;192:132–141.

98   Schlesinger JJ, Chapman S: Neutralizing F(ab′)2 fragments of protective monclonal antibodies to yellow fever virus (YF) envelope protein fails to protect mice against lethal YF encephalitis. J Gen Virol 1995;76:217–220.

99   McCullough KC, Crowther JR, Butcher RN, Carpenter WC, Brocchi E, Cappucci L, De Simone F: Immune protection against foot-and-mouth disease virus studied using neutralizing and non-neutralizing concentrations of monoclonal antibodies. Immunology 1986;58:421–428.

100   Boere WAM, Benaissa-Trouw BJ, Harmsen T, Erich T, Kraaijeveld CA, Snippe H: Mechanisms of monoclonal antibody-mediated protection against virulent Semliki Forest virus. J Virol 1985;54:546–551.

101   Prince GA, Hemming VG, Horswood RL, Baron PA, Murphy BR, Chanock RM: Mechanism of antibody-mediated clearance in immunotherapy of respiratory syncytial virus infection of cotton rats. J Virol 1990;64:3091–3092.

102   Mariani G, Strober W: Immunoglobulin metabolism; in Metzger H (ed): Fc Receptors and the Action of Antibodies. Washington, American Society for Microbiology, 1990.

103   Crowe JE Jr, Murphy BR, Chanock RM, Williamson RA, Barbas CF III, Burton DR: Recombinant human respiratory syncytial virus (RSV) monoclonal antibody Fab is effective therapeutically when introduced directly into the lungs of RSV-infected mice. Proc Natl Acad Sci USA 1994;91: 1386–1390.

104   Parren PWHI, Ditzel HJ, Gulizia RJ, Binley JM, Barbas CF III, Burton DR, Mosier DE: Protection against human immunodeficiency virus type 1 infection in hu-PBL-SCID mice by passive immunization with a neutralizing human monoclonal antibody against the gp120 CD4-binding site. AIDS 1995;9:F1–F6.

105   Robey WG, Arthur LO, Matthews TJ, Langlois A, Copeland TD, Lerche NW, Oroszlan S, Bolognesi DP, Gilden RV, Fischinger PJ: Prospect for prevention of human immunodeficiency virus infection: Purified 120 kDa envelope glycoprotein induces neutralizing antibody. Proc Natl Acad Sci USA 1986;83:7023–7027.

106   Javaherian K, Langlois AJ, McDanal C, Ross KL, Eckler LI, Jellis CL, Profy AT, Rusche JR, Bolognesi DP, Putney SD, Matthews TJ: Principal neutralizing domain of the human immunodeficiency virus type 1 envelope protein. Proc Natl Acad Sci USA 1989;86:6768–6772.

107   Steimer KS, Scandella CJ, Skiles PV, Haigwood NL: Neutralization of divergent HIV-1 isolates by conformation-dependent human antibodies to gp120. Science 1991;254:105–108.

108   Moore JP, Sattentau QJ, Wyatt R, Sodroski J: Probing the structure of the human immunodeficiency virus surface glycoprotein gp120 with a panel of monoclonal antibodies. J Virol 1994;68:469–484.

109   Moore JP, Ho DD: HIV-1 neutralization: The consequences of viral adaptation to growth on transformed T cells. AIDS 1995;9(suppl A):S117–S136.

110   Javaherian K, Langlois AJ, LaRosa GJ, Profy AT, Bolgnesi DP, Herlihy WC, Putney SD, Matthews TJ: Broadly neutralizing antibodies elicited by the hypervariable neutralizing determinant of HIV-1. Science 1990;250:1590–1593.

111   LaRosa GJ, Davide JP, Weinhold K, Waterbury JA, Profy AT, Lewis JA, Langlois AJ, Dreesman GR, Boswell RN, Shadduck P, Holley LH, Karplus M, Bolognesi DP, Matthews TJ, Emini EA, Putney SD: Conserved sequence and structural elements in the HIV-1 principal neutralizing determinant. Science 1990;249:932–935.

112   Moore JP, Nara PL: The role of the V3 loop in HIV infection. AIDS 1991;5(suppl 2):S21–S33.

113 Moore JP, Ho DD: Antibodies to discontinuous or conformationally-sensitive epitopes on the gp120 glycoprotein of human immunodeficiency virus type 1 are highly prevalent in sera of infected humans. J Virol 1993;67:863–875.

114 Ho DD, McKeating JA, Li XL, Moudgil T, Daar ES, Sun N-C, Robinson JE: Conformational epitope on gp120 important in CD4 binding and human immunodeficiency virus type 1 neutralization identified by a human monoclonal antibody. J Virol 1991;65:489–493.

115 Tilley SA, Honnen WJ, Racho ME, Hilgartner M, Pinter A: A human monoclonal antibody against the CD4-binding site of HIV1 gp120 exhibits potent, broadly neutralizing activity. Res Virol 1991; 142:247–259.

116 Thali M, Moore JP, Furman C, Charles M, Ho DD, Robinson J, Sodroski J: Characterization of conserved human immunodeficiency virus type 1 gp120 neutralization epitopes exposed upon gp120-CD4 binding. J Virol 1993;67:3978–3988.

117 Gorny MK, Moore JP, Conley AJ, Karwowska S, Sodroski J, Williams C, Burda S, Boots LJ, Zolla-Pazner S: Human anti-V2 monoclonal antibody that neutralizes primary but not laboratory isolates of human immunodeficiency virus type 1. J Virol 1994;68:8312–8320.

118 Trkola A, Purtscher M, Muster T, Ballaun C, Buchacher A, Sullivan N, Srinivasan K, Sodroski J, Moore JP, Katinger H: Human monoclonal antibody 2G12 defines a distinctive neutralization epitope on the gp120 glycoprotein of human immunodeficiency virus type I. J Virol 1996;70:1100–1108.

119 Pincus SH, Messer KG, Nara PL, Blattner A, Colclough G, Reitz M: Temporal analysis of the antibody response to HIV envelope protein in HIV-infected laboratory workers. J Clin Invest 1994; 93:2505–2513.

120 Muster T, Guinea R, Trkola A, Purtsher M, Klima A, Steindl F, Palese P: Cross-neutralizing activity against divergent human immunodeficiency virus type 1 isolates induced by the gp41 sequence ELDKWAS. J Virol 1994;68:4031–4034.

121 Conley AJ, Kessler JA II, Boots LJ, Tung J-S, Arnold BA, Keller PM, Shaw AR, Emini EA: Neutralization of divergent human immunodeficiency type 1 variants and primary isolates by IAM-41-2F5, and anti-gp41 human monoclonal antibody. Proc Natl Acad Sci USA 1994;91:3348–3352.

122 Daar ES, Li XL, Moudgil T, Ho DD: High concentrations of recombinant soluble CD4 are required to neutralize primary human immunodeficiency virus type 1 isolates. Proc Natl Acad Sci USA 1990;87:6574–6578.

123 Bou-Habib DC, Roderiquez G, Oravecz T, Berman PW, Lusso P, Norcross MA: Cryptic nature of envelope V3 region epitopes protects primary human immunodeficiency virus type 1 from antibody neutralization. J Virol 1994;68:6006–6013.

124 Burton DR, Barbas CF III, Persson MA, Koenig S, Chanock RM, Lerner RA: A large array of human monoclonal antibodies to type 1 human immunodeficiency virus from combinatorial libraries of asymptomatic seropositive individuals. Proc Natl Acad Sci USA 1991;88:10134–10137.

125 Barbas CF III, Collet TA, Amberg W, Roben P, Binley JM, Hoekstra D, Cababa D, Jones TM, Williamson RA, Pilkington GR, Haigwood NL, Cabezas E, Satterthwait AC, Sanz I, Burton DR: Molecular profile of an antibody response to HIV-1 as probed by combinatorial libraries. J Mol Biol 1993;230:812–823.

126 Moore JP, Ho DD: Antibodies to discontinuous or conformationally sensitive epitopes on the gp120 glycoprotein of human immunodeficiency virus type 1 are highly prevalent in sera of infected humans. J Virol 1993;67:863–875.

127 Barbas CF III, Bjorling E, Chiodi F, Dunlop N, Cababa D, Jones TM, Zebedee SL, Persson MA, Nara PL, Norrby E, Burton DR: Recombinant human Fab fragments neutralize human type 1 immunodeficiency virus in vitro. Proc Natl Acad Sci USA 1992;89:9339–9343.

128 Helseth E, Kowalski M, Gabuzda D, Olshevsky U, Haseltine W, Sodroski J: Rapid complementation assays measuring replicative potential of human immunodeficiency virus type 1 envelope glycoprotein mutants. J Virol 1990;64:2416–2420.

129 Hanson CV, Crawford-Miksza L, Sheppard HW: Application of a rapid microplaque assay for determination of human immunodeficiency virus neutralizing antibody titers. J Clin Microbiol 1990; 28:2030–2034.

130   Trkola A, Pomales AP, Yuan H, Korber B, Maddon PJ, Allaway G, Katinger H, Barbas CF III, Burton DR, Ho DD, Moore JP: Cross-clade neutralization of primary isolates of human immunodeficiency virus type 1 by human monoclonal antibodies and tetrameric CD4-IgG. J Virol 1995;69:6609–6617.

131   Gauduin MC, Allaway GP, Maddon PJ, Barbas CF III, Burton DR, Koup RA: Effective ex vivo neutralization of plasma HIV-1 by recombinant immunoglobulin molecules. J Virol, in press.

132   Torbett BE, Picchio G, Mosier DE: Hu-PBL-SCID mice: A model for human immune function. AIDS, and lymphoma genesis. Immunol Rev 1991;124:139–164.

133   Hesselton RM, Koup RA, Cromwell MA, Graham BS, Johns M, Sullivan JL: Human peripheral blood xenografts in the SCID mouse: Characterization of immunologic reconstitution. J Infect Dis 1993;168:630–640.

134   Koup TA, Hesselton RM, Safrit JT, Somasundran M, Sullivan JL: Quantitative assessment of human immunodeficiency virus type 1 replication in human xenografts of acute-infected Hu-PBL-SCID mice. AIDS Res Hum Retroviruses 1994;10:279–284.

135   Zuckier LS, Rodriguez LD, Scharff MD: Immunologic and pharmacologic concepts of monoclonal antibodies. Semin Nucl Med 1989;19:166–186.

136   Ditzel HJ, Parren PWHI, Binley JM, Sodroski J, Moore JP, Barbas CF III, Burton DR: Mapping the protein surface of HIV-1 gp120 using human monoclonal antibody libraries from phage display libraries, unpubl.

137   Ditzel HJ, Binley JM, Moore JP, Sodroski J, Sullivan N, Sawyer LSW, Hendry RM, Yang W-P, Barbas CF III, Burton DR: Neutralizing recombinant human antibodies to a conformational V2- and CD4-binding site-sensitive epitope of HIV-1 gp120 isolated using an epitope-masking procedure. J Immunol 1995;154:895–908.

138   Helseth E, Olshevsky U, Furman C, Sodroski J: Human immunodeficiency virus type 1 gp120 envelope glycoprotein regions important for association with the gp41 transmembrane glycoprotein. J Virol 1991;65:2119–2123.

139   Olshevsky U, Helseth E, Furman C, Li J, Haseltine W, Sodroski J: Identification of individual human immunodeficiency virus type 1 gp120 amino acids important for CD4 receptor binding. J Virol 1990;64:5701–5707.

140   Ditzel HJ, Parren PWHI, Sullivan N, Sodroski J, Sawyer LSW, Hendry RM, Pinter A, Dunlop N, Nara PL, Trkola A, Moore JP, Barbas CF III, Burton DR: A recombinant human Fab fragment directed against a conformation dependent epitope in the second variable region of HIV-1 gp120 neutralizes T-cell line-adapted strains of HIV-1 but primary isolates less efficiently, in preparation.

141   Ho DD, Neumann AU, Perelson AS, Chen W, Leonard JM, Markowitz M: Rapid turnover of plasma virions and CD4 lymphocytes in HIV-1 infection. Nature 1995;373:123–126.

142   Wei X, Ghosh SK, Taylor ME, Johnson VA, Emini EA, Deutsch P, Lifson JD, Bonhoeffer S, Nowak MA, Hahn BH, Saag MS, Shaw GM: Viral dynamics in human immunodeficiency virus type 1 infection. Nature 1995;373:117–122.

143   Binley JM, Ditzel HJ, Barbas CF III, Sullivan N, Sodroski J, Parren PWHI, Burton DR: Human antibody responses to HIV-1 gp41 cloned in phage display libraries suggest three major epitopes are recognized and give evidence for conserved antibody motifs in antigen binding. AIDS Res Hum Retroviruses, in press.

144   Roben P, Moore JP, Thali M, Sodroski J, Barbas CF III, Burton DR: Recognition properties of a panel of human recombinant Fab fragments to the CD4 binding site of gp120 that show differing abilities to neutralize human immunodeficiency virus type 1. J Virol 1994;68:4821–4828.

145   Chothia C, Lesk AM, Gherardi E, Tomlinson IM, Walter G, Marks JD, Llewelyn MB, Winter G: Structural repertoire of the human V_H segments. J Mol Biol 1992;227:799–817.

146   Sattentau QJ, Zolla-Pazner S, Poignard P: Epitope exposure on functional, olgomeric HIV-1 gp41 Molecules. Virology1995;206:713–717.

147   Barbas CF III, Hu D, Dunlop N, Sawyer L, Cababa D, Hendry RM, Nara PL, Burton DR: In vitro evolution of a neutralizing human antibody to human immunodeficiency virus type 1 to enhance affinity and broaden strain cross-reactivity. Proc Natl Acad Sci USA 1994;91:3809–3813.

148 Yang W-P, Green K, Pinz-Sweeney S, Briones AT, Burton DR, Barbas CF III: CDR walking mutagenesis for the affinity maturation of a potent human anti-HIV-1 antibody into the picomolar range. J Mol Biol 1995;254:392–403.

149 Sattentau Q, Parren PW, Burton DR: Unpubl.

150 Earl, PL, Broder CC, Long D, Lee SA, Peterson J, Chakrabarti S, Doms RW, Moss B: Native oligomeric human immunodeficiency virus type 1 envelope glycoprotein elicits diverse monoclonal antibody reactivities. J Virol 1994;68:3015–3026.

151 Broder CC, Earl PL, Long D, Abedon ST, Moss B, Doms RW: Antigenic implications of human immunodeficiency virus type 1 envelope quaternary structure: Oligomer-specific and -sensitive monoclonal antibodies. Proc Natl Acad Sci USA 1994;91:11699–11703.

152 Parren PWHI, Fisicaro P, Labrijn A, Binley JM, Yang W-P, Ditzel HJ, Barbas CF III, Burton DR: In vitro antigen 'challenge' of human antibody libraries for vaccine evaluation: The human immunodeficiency virus type 1 oligomer. J Virol 1996, in press.

153 Buchacher A, Predl R, Strutzenberger K, Steinfellner W, Trkola A, Purtscher M, Gruber G, Tauer C, Steidl F, Jungbauder A, Katinger H: Generation of human monoclonal antibodies against HIV-1 proteins; Electrofusion and Epstein-Barr virus transformation for peripheral blood lymphocyte immortalization. AIDS Res Hum Retroviruses 1994;10:359–369.

154 Burton DR, Barbas CF III: Human antibodies to HIV-1 by recombinant DNA methods; in Norrby E (ed): Immunochemistry of AIDS. Chem Immunol Basel, Karger, 1993, vol 56, pp 112–126.

155 Vogt A, Kopp R, Maass G, Reich L: Poliovirus type 1: Neutralization by papain-digested antibodies. Science 1964;145:1447–1448.

156 Klinman NR, Long CA, Karush F: The role of antibody bivalence in the neutralization of bacteriophage. J Immunol 1967;99:1128–1133.

157 Ashe WK, Mage M, Mage R, Notkins AL: Neutralization and sensitization of herpes simplex virus with antibody fragments from rabbits of different allotypes. J Immunol 1968;101:500–504.

158 Rosenstein RW, Nisonoff A, Urh JW: Significance of bivalence of antibody in viral neutralization. J Exp Med 1971;134:1431–1441.

159 Merz DC, Scheid A, Choppin PW: Immunological studies of the functions of paramyxovirus glycoproteins. Virology 1981;109:94–105.

160 Roehrig JT, Hunt AR, Kinney RM, Mathews JH: In vitro mechanisms of monoclonal antibody neutralization of alphavirus. Virology 1988;165:66–73.

161 Colonno RJ, Callahan PL, Leippe DM, Rueckert RR, Tomassini JE: Inhibition of rhinovirus attachment by neutralizing monoclonal antibodies and their Fab fragments. J Virol 1989;63:36–42.

162 Kingsford L, Boucquey KH, Porter T, Cardoso TP: Effects of specific monoclonal antibodies on la crosse virus neutralization: Aggregation, inactivation by fab fragments, and inhibition of attachment to baby hamster kidney cells. Virology 1991;180:591–601.

163 Cheung SC, Dietzschold B, Koprowski H, Notkins AL, Rando RF: A recombinant human Fab expressed in escherichia coli neutralizes rabies virus. J Virol 1992;66:6714–6720.

Dr. Paul W.H.I. Parren, The Scripps Research Institute, Department of Immunology, 10550 North Torrey Pines Road (Imm 2), La Jolla, CA 92037 (USA)

Capra JD (ed): Antibody Engineering.
Chem Immunol. Basel, Karger, 1997, vol 65, pp 57–72

....

Chemical Engineering at the Antibody Hinge

George T. Stevenson

Tenovus Research Laboratory, Southampton University Hospitals, Southampton, UK

The hinge region in IgG antibodies presents a cluster of disulfide (SS) bonds which may be reduced to yield sulfhydryl (SH) groups at known positions. These groups, the most reactive to be found in proteins, can be manipulated by SS interchange and alkylation to build up mosaic constructs from modules such as Fab'γ and Fcγ in a wide variety of configurations [1]. The products have been intended mainly for therapeutic use, particularly the destruction of neoplastic cells, but they can be designed also for other purposes such as immunochemical assays and immunohistology.

The technology described here can complement the much more widely practised genetic engineering of immunoglobulin molecules in two major ways: by providing a rapid prediction of performance of constructs such as bispecific antibodies before investing in a genetic synthesis, and by offering many other antibody configurations which would be very difficult to achieve by genetic means. Clearly, there is great potential in combining the two technologies, say by chemically engineering a recombinant Fab' which has had cysteine residues inserted at predetermined positions [2]. However, the problem of availability of expression systems which can economically provide large amounts of product, such as are needed for clinical use, means that most chemical engineering is still practised on monoclonal antibodies and their fragments.

Essential Anatomy of the Antibody Molecule

The two heavy (γ) and two light (κ or λ) chains of an IgG molecule are held together by sets of noncovalent and SS bonds (fig. 1). Cleavage of the SS bonds leaves the four-chain structure still secured by its noncovalent interchain bonds. However, the lack of significant horizontal interactions between the two hinges, and between the two Cγ2 domains, means that the hinge is free to open out after loss of the inter-γ SS bonds: the Y-shaped molecular envelope seen on electron microscopy of antigen-antibody complexes is altered so as to suggest a C-terminal transposition of the hinge to the junction of the Cγ2 and Cγ3 domains [3]. There is an accompanying loss of ability to activate the classical complement pathway [4]. In contrast there is no discernible loss of antibody association constant, at least for haptenic epitopes, following cleavage of heavy-light SS bonds [5, 6].

The Immunoglobulin Hinge

The *genetic hinge* is a cysteine- and proline-rich stretch of the γ-chain which, in the prototypic molecule human IgG1, runs from residues 216 to 230 (Eu numbering). As its name implies, it is encoded by its own exon. It overlaps the *structural hinge*, which is defined by crystallographic evidence of molecular mobility and runs from residues 221 to 237. The structural hinge in fact functions as two hinges, one at either end of the rigid SS-bonded sequence (fig. 1). The upper hinge is believed to permit both wagging and rotation of each Fab arm, while the lower hinge permits wagging of the Fc [7].

Enzymatic cleavages at the hinge, first reported more than 30 years ago, remain one of the day-to-day tools of the immunochemist. Despite some variation, there is a broad resemblance in susceptibility to enzymatic cleavage shown by IgGs from various species, and by the IgG subclasses and allotypes within a species. As shown in figure 1, papain cleaves the γ-chains of human IgG1 just N-terminal to the inter-γ SS bonds, to yield one Fcγ1 and two Fabγ fragments. At a pH of ∼4, some 1.5 units above its optimum, pepsin cuts the γ-chains on the C-terminal side of the inter-γ SS bonds and degrades the Cγ2 domain so that only fragments of Fc survive. We are left with the dimeric $F(ab'γ)_2$, the prime being used to denote the persistence of most of the hinge on the Fab arms. Reduction of the inter-γ SS bonds with a thiol leads the two Fab'γ to separate.

Mouse IgG1 and IgG2a antibodies, the two isotypes most commonly yielded by standard monoclonal protocols, are readily digested by pepsin to yield $F(ab')_2$ as described above [8]. Each of these isotypes has three inter-γ SS bonds, compatible with the requirement of our techniques for an odd number of such bonds in Fab' modules. (It will be seen later that this requirement arises from the tendency of two neighboring SH groups either to disulfide-bond, or to form intrachain loops with crosslinkers such as bismaleimides.) IgG2a of non-BALB/c origin is sometimes present as a 'B' allotype with only two inter-γ SS bonds, while IgG2b (which yields $F(ab')_2$ upon digestion by

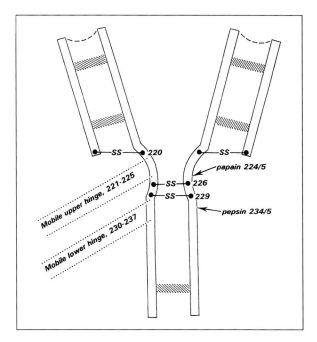

Fig. 1. Interchain bonds and the hinge region in human IgG1. Numbers refer to amino acid residues, using the Eu system. Cross-hatching indicates sets of noncovalent bonds linking immunoglobulin domains: from top to bottom, V_L to V_H, C_L to $C\gamma1$, $C\gamma3$ to $C\gamma3$. Sites of mobility define the functional upper and lower hinges, which overlap the genetic hinge (residues 216 to 230). Arrows indicate the sites of enzymic cleavage.

lysyl endopeptidase [9]) has four: use of Fab' from either of these would require significant modifications to the methods described in this chapter.

Functional Types of SS Bonds

The IgG molecule illustrates the fact that protein SS bonds are divisible into three functional types with regard to manipulation by SS interchange with thiols: shielded bonds (usually intrachain), assisted bonds and unassisted bonds (both usually interchain). In figure 2 this concept is illustrated by the SS bonds in a dimeric F(ab')$_2$ fragment.

In the absence of gross unfolding of the protein by a denaturing agent, *shielded* bonds are simply inaccessible to thiols. The intrachain SS bond which closes the immunoglobulin domain loop is such a bond: it links the two β-pleated sheets of the domain, and is immersed in the hydrophobic interface between them.

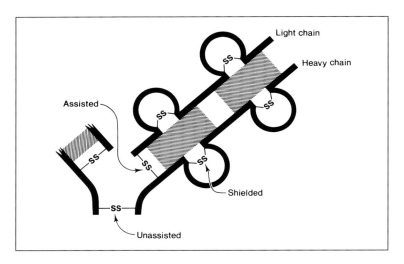

Fig. 2. Three differing types of SS bonds in an F(ab')$_2$ fragment. Only one inter-γ chain bond is depicted, as occurs in rabbit F(ab'γ)$_2$: but even multiple such bonds, offering mutual reinforcement, remain unassisted as regards the localizing of half-cystine residues by noncovalent structure.

Assisted SS bonds can be attacked by thiols, but after reduction they yield two cysteine residues still held in close proximity by noncovalent bonds within the protein molecule. This proximity favors reformation of the SS bond, either by SS interchange or by oxidation. An *unassisted* SS bond lacks such noncovalent back-up, so that reduction leaves its two cysteine residues free to move apart. The solitary inter-γ chain bond at the F(ab')$_2$ hinge shown in figure 2 is such a case. The presence of multiple inter-γ SS bonds, as occur in the mouse F(ab')$_2$ which we deal with in this chapter, offers some mutual reinforcement; but once all are broken the homologous cysteine residues will drift apart. We have found it possible to take considerable advantage of the difference between assisted and unassisted bonds to accomplish selective SS interchange.

Manipulating Cysteine Residues

A comprehensive review of the sulfur chemistry of proteins has been provided by Liu [10]. In the manipulations to be described here only two major types of reaction are utilized: (1) SS interchange involving attack by a thiol on a disulfide and (2) alkylation of SH groups by maleimides. While carrying out these reactions one should constantly be aware of possible problems from an inadvertent third reaction, oxidation of SH groups to disulfides by oxygen.

SS interchange

The general equation for SS interchange – the reduction of a disulfide (P-SS-Q) by a thiol (R-SH)[1] – is frequently written:

P-SS-Q + R-SH $\rightleftharpoons$ P-SS-R + Q-SH

The reaction is strictly a nucleophilic attack by the thiol ion R-S$^-$, with the pH dependence of the rate reflecting the fact that most thiols exhibit a pK of about 9. Progress of the reaction eventually gives R-SH, P-SH and Q-SH each reacting with R-SS-R, P-SS-P, Q-SS-Q, R-SS-P, P-SS-Q and Q-SS-R: 18 rate constants are required to describe the equilibrium, and frequently an over-looked product turns up to cause some sort of problem. However, several well-understood factors which alter the equilibrium can be readily manipulated. By mass action, a sufficient surplus of R-SH in the above equation will lead to near-total conversion of P-SS-Q to P-SH and Q-SH:

P-SS-Q + 2R-SH $\rightarrow$ P-SH + Q-SH + R-SS-R

Certain dithiols, of which dithiothreitol (DTT) is a well-known example [12], will drive the reaction to the right without being in large molar excess, because of their tendency to form an intramolecular SS bond:

P-SS-Q + threitol(-SH)$_2$ $\rightarrow$ P-SH + Q-SH + threitol(-SS-).

Another interesting and important example of essentially one-way SS interchange is provided by compounds with SS bonds attached to the 2 or 4 carbon of a pyridyl ring, such as 4,4'-dipyridyl disulfide (Py-SS-Py) [13–15]. Reduction releases the pyridylthiol which, instead of reducing a disulfide in its turn, is almost entirely removed by a one-sided tautomeric equilibrium with the thiopyridone (Py=S):

Another important property of the dipyridyl disulfides, and seen to a variable extent in mixed pyridyl disulfides such as protein-SS-Py, is due to protonation of the pyridyl nitrogen rendering the SS bond unusually electrophilic at low pH. To appreciate normal reactivity of protein SS bonds, note

---

[1] The term 'SS interchange' was applied originally to reactions of the type P-SS-P + Q-SS-Q $\rightarrow$ 2P-SS-Q. When occurring at near-neutral pH, these are initiated by small amounts of a thiol such as P-SH and then funnel through thiol-disulfide interactions. A similar overall SS interchange occurring in strongly acid solution has a different mechanism, and is of interest in having caused problems almost 50 years ago in locating the SS bonds in insulin. See an early review by Cecil and McPhee [11].

that human IgG1, with four interchain SS bonds, yields an SH titer of $<0.2/$ molecule after exposure to 1 mM DTT at pH 5.0 for 30 min at 25 °C. This result reflects the low reactivity of the highly protonated DTT. In contrast, the SS bond in Fab'-SS-Py (whose preparation we describe later) is essentially entirely reduced by the same treatment to Fab'-SH, the reaction apparently funnelling rapidly through the minute proportion of ionized DTT present. It is instructive to summarize the reactions involved:

(a) Fab'-SS-Py + threitol(-SH)$_2$ → Fab'-SS-threitol-SH + Py=S
(b) Fab'-SS-threitol-SH → Fab'-SH + threitol(-SS-).

Reaction (a) is promoted by the electrophilic nature of the pydidyl-linked SS bond, and rendered unidirectional by the tautomeric formation of Py=S and by the occurrence of reaction (b). A strong chromophoric signal from Py=S ($\varepsilon_{mM} = 19.8$ at pH 7.2) provides a ready assessment of the extent of reaction. Reaction (b) reflects the strong tendency of DTT to cyclize through an SS bond by intramolecular SS interchange, even at pH 5.

Important roles played by pyridyl disulfides are illustrated later in accounts of preparing Fab'γ and Fcγ modules.

Steric effects on SS interchange include impeded access of thiols to some SS bonds, and promotion of SS formation when two SH groups are held in close proximity. Both of these factors have been alluded to in our discussion of SS bonds in immunoglobulin molecules and reappear when describing preparations of the modules.

It may be useful to summarize here the devices used in controlling SS interchange: (1) mass action (surplus of a reactant, or continuous removal of a product); (2) vicinal charge; (3) pH, which of course affects vicinal charge; (4) steric effects.

Alkylation by Maleimides
Alkylation of SH groups is used either for blocking them, or for adding a cross-linking reagent. For simple blocking iodoacetate or iodoacetamide is often used. However, greater selectivity for SH groups is available from maleimides, such as N-ethylmaleimide (NEM), where alkylation involves an addition to the double bond:

Again the reaction involves the thiolate ion R-S$^-$, so the rate falls with decreasing pH. However, we find that the rate at pH 5.0 is adequate, both for NEM and for bismaleimide linkers, and working at this relatively low pH is useful in hindering unwanted SS interchange, oxidation of SH groups, and hydrolysis of maleimide groups.

Bismaleimide linkers such as o-phenylenedimaleimide (PDM) offer a highly satisfactory means of linking SH-displaying protein modules. First one module reacts with the linker, which is present at a sufficient concentration to avoid homologous cross-linking of the module. One obtains protein-S-(o-succinimidylphenyl)-maleimide. After separation from surplus linker, the protein-maleimide is allowed to react with the second protein partner to yield protein-S-R-S-protein, where the intermodular link consists of tandem thioether bonds separated by the R group (in this case o-phenylene-disuccini-midyl). This tandem thioether linkage has proved to be stable in vivo [16].

During the reaction of PDM with a first partner displaying more than one SH group, there will be a strong tendency to form intramolecular cross-links. This need not cause any problem if the molecule presents an odd number of SH groups and one wants it finally to present only a single maleimide group. For example, Fab'(-SH)$_5$ plus PDM yields Fab'(-maleimide)$_1$ [17]. If starting with an even number of SH, closely spaced like those resulting from reduction of the interchain SS bonds of immunoglobulin modules, reaction with PDM is apt to yield no active maleimide group attached to the protein. One must then start thinking of solutions such as trimaleimide linkers.

Oxidation of SH Groups

It is surprising how often conversion of SH groups to SS bonds by SS interchange is confused with oxidation by molecular oxygen:

$$R\text{-}S^- + R'\text{-}S^- \rightarrow R\text{-}SS\text{-}R' + 2e.$$

Given two SH groups sterically able to form an SS bond, such oxidation represents a simple and commonly practised way of having them do so: one need only incubate the solution at a pH suitable for ionization of SH to S$^-$ (say >8.0), in the presence of oxygen. The major problem is that the rate is not predictable and reproducible in the manner of SS interchange: normally the electron transfer appears to proceed via metal catalysts (especially Fe^{3+} and Cu^{2+}) but the mechanisms are complex. Our approach to oxidation is to avoid it. This one does by observing standard precautions for handling SH groups: keep the pH and temperature low, have a chelating agent such as EDTA present, and exclude oxygen by having the buffers N$_2$-saturated.

The Fab'γ Module

F(ab')₂

We obtain F(ab')₂ by limited proteolysis of murine IgG1 or IgG2a antibody (14mg/ml) with pepsin (1 mg/ml), the digestion usually taking place at pH 3.9 for about 6 h. Minor variations in susceptibility to digestion are encountered which are not explicable in terms of antibody sequence but might reflect different degrees of glycosylation. These variations tend to be reproducible from one sample to another of a given protein and are dealt with by varying the pH by up to ± 0.2 unit: in our experience, the control of pH is quite critical. The progress of digestion is followed at hourly intervals by HPLC, and the digestion halted when about 90% complete by raising the pH to 7.7. Attempts to pursue the digestion to completion are apt to be accompanied by a disproportionate degradation of the F(ab')₂. The F(ab')₂ is separated from IgG, small digestion fragments and enzyme by recycling gel chromatography on Superdex 200 (Pharmacia). Yields of 60–70% are not unusual.

Fab'(SH)₅

A simple reduction of the F(ab')₂ (5 mg/ml) with DTT (1 mM), at pH 8.0, yields Fab'(-SH)₅. The protein is separated from the DTT, and at the same time transferred to acetate buffer of pH 5.0, by passage through Sephadex G-25 (Pharmacia). For some purposes, Fab'(-SH)₅ is a suitable module for conjugation, for example in the preparation of trispecific Fab₃ or the chimeric construct FabFc₂² [1, 17]. Although Fab'(-SH)₅ is theoretically pentavalent for reaction with maleimides, attempts to attach more than three Fc-maleimide modules evidently begin to encounter steric hindrance. Up to this point, the ratio of Fc-maleimide to Fab'(-SH)₅ in the reaction mixture determines the predominant species to be formed, so that a ratio of 2:1 yields predominantly FabFc₂. Products of such reactions can be separated by gel chromatography on Superdex 200. If in the final preparation one need deliver only an average stoichiometry of FabFc₂ in a range from FabFc to FabFc₃, instead of monodisperse FabFc₂, the recovery of Fab' doubles to about 70%.

Fab'-SS-Py

Most of our current use of Fab' modules requires *univalent* Fab'-SH. This is used particularly in the preparation of bispecific F(ab')₂ which retains potential hinge-region SH groups for addition of further modules. The Fab'-SH is reached from Fab'(-SH)₅ through two further stages of SS interchange, depicted in figure 3.

---

² To avoid too cumbersome a nomenclature, primes and Greek letters are omitted from the names of final constructs.

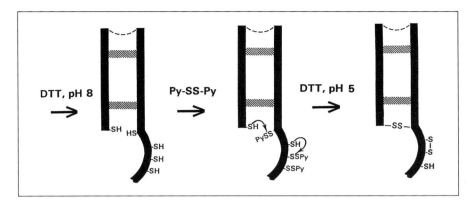

Fig. 3. Disulfide interchanges which convert F(ab')$_2$, derived from murine IgG1 or IgG2a antibody, to Fab'-SH. An initial reduction by DTT at pH 8 cleaves the dimer to give Fab'(-SH)$_5$. Reaction with Py-SS-Py at a fairly arbitrary pH (say 4–9) leads to Fab'-SS-Py, after four of the SH groups form two intramolecular SS bonds by intramolecular attacks on transiently formed mixed pyridyl disulfides. We do not know which two of the three candidate SH groups form the longitudinal SS bond in the hinge: the depiction of the N-terminal pair is arbitrary. Finally, a limited attack by DTT at pH 5 breaks the electrophilic pyridyl-linked SS bond, and this bond only.

In the first SS interchange, Py-SS-Py is added to Fab'(SH)$_5$ to form Fab'-SS-Py, an intermediate which can serve as a storage form of univalent Fab'. The pH for the interchange has not proved critical, but we have chosen 5.0 rather than 8.0 because the slower reaction at pH 5 is likely to pose fewer problems arising from transient concentrations during mixing. Fab'(SH)$_5$ is obtained at ∼3 mg/ml in acetate buffer of pH 5.0 as described above. The protein is now kept cold and under N$_2$ while stirring, and Py-SS-Py (15 m*M* in 50% v/v dimethylformamide/acetate buffer) is added over a period of 1 min to a final concentration of 1 m*M*. After separation of the Fab' on Sephadex, titration reveals 0.85–1.1 SS-Py group per molecule. Even with abrupt addition of Py-SS-Py to a final concentration as high as 5 m*M* we have rarely found >1.5 group per molecule, in contrast to 5 groups expected from complete conversion of SH to SS-Py. An explanation for this phenomenon is seen in figure 3: an initial formation of Fab'-SS-Py is apt to be followed, whenever a nearby SH exists, by an SS interchange which forms an intramolecular SS bond more rapidly than the same nearby SH reacts with Py-SS-Py. By this means, the γ light SS bond is reconstituted, as was expected. However, unexpectedly, an intrachain SS bond not seen in the native protein forms within the hinge to leave only one cysteine residue bearing an SS-Py group. Formation of such an SS-bonded loop has also been reported when the SH groups in mouse Fab′γ are allowed to undergo spontaneous oxidation [18].

We store Fab′ modules in the form Fab′-SS-Py at 5 °C for up to one week, or deep frozen for longer periods. The SS-Py group appears to be easily reduced by contaminants: plastic storage vessels should be avoided, and it might be advisable to store in the presence of surplus Py-SS-Py. All chromatography of Py-SS-Py, and of protein displaying SS-Py groups, should be conducted in the presence of a subdenaturing concentration of organic solvent in order to prevent the pyridyl group adhering to the column, particularly to plastic end-pieces; we have found 1.0 M (7.7% v/v) dimethylformamide suitable.

Fab′-SH and Fab′-Maleimide

These are the reactive forms of the Fab′γ module, to be used immediately for conjugation. After separation from Py-SS-Py, Fab′-SS-Py is allowed to react with 1 mM DTT at pH 5.0 for 30 min at 25 °C to yield, as described previously, univalent Fab′-SH which can be separated by chromatography on Sephadex G-25.

Depending upon the plan of construction, the Fab′-SH may be conjugated in this form, or may first be converted to Fab′-maleimide by reaction with the PDM linker: to Fab′-SH, in N$_2$-saturated buffer at pH 5.0, add 0.2 times its volume of 6 mM PDM in 50% dimethylformamide, incubate at 5 °C for 30 min, then separate the resulting Fab′-maleimide on Sephadex G-25.

The Fcγ Module

Fcγ1 from Basic IgG

Despite the potential attractions of genetically engineered human Fcγ in some sort of unicellular expression system, IgG from human plasma remains by far the most convenient source for large quantities of Fcγ. A simple calculation reveals that a donation of a single unit of blood (450 ml) by every adult in the USA would provide approximately 500 tons of IgG. Some 70% of this plasma IgG belongs to the subclass IgG1 whose Fc module we require. Because the processing of plasma is at present driven by requirements for albumin and coagulation factor VIII, surplus IgG can often be purchased relatively cheaply in the form of fraction II precipitate yielded by the Cohn cold ethanol process. Local safety requirements for handling human blood products (regarding especially HIV, and the hepatitis B and hepatitis C viruses) must of course be taken into account.

Starting with fraction II paste, we remove the most acidic third of the protein by ion-exchange chromatography on DEAE-Sepharose (Pharmacia).

The resulting *basic IgG* is freed of non-immunoglobulin impurities, digests more uniformly with papain than does total IgG, and offers ready chromatographic separation of Fcγ from the notably more basic Fabγ and residual IgG. Our conditions for digestion and purification yield essentially only the Fc from IgG1, i.e. Fcγ1: the only other class detectable by radial immunodiffusion has been Fcγ2, at <1% the concentration of Fcγ1.

To minimize damage to the Fc, digestion of basic IgG by papain is carried out under the minimal conditions required for >80% cleavage of the IgG1 hinge: IgG 18 mg/ml, papain 0.05 mg/ml, 2-mercaptoethanol 5 mM, pH 6.7, 37 °C for 20 min. Digestion is terminated by adding NEM to 2 mM, and the Fcγ1 separated by ion-exchange chromatography. The concentration of 2-mercaptoethanol used to maintain papain in its SH form is too low to cause any detectable cleavage of the IgG1 hinge SS, an important point because all SH groups in the digest solution will be blocked permanently when NEM is added. From individual lots of 12 g of basic IgG we harvest about 2.7 g of Fcγ1.

Fc-SS-Py

The usual form in which the Fc module is used for conjugation to another protein is *Fc-maleimide*, in which a single maleimide group protrudes from a cysteine residue in the hinge. The inter-γ SS bond in which this cysteine participated is necessarily sacrificed. The remaining SS bond remains intact, so preserving the ability to activate complement, but it can readily be left open – if for example it is considered that activation of complement might add to the toxicity of a therapeutic conjugate.

Due to instability of the maleimide group, we again find it convenient to store the module at the intermediate stage of its preparation where the linking cysteine exists as a pyridyl disulfide, that is in the form Fc-SS-Py. The various stages in the preparation are shown in figure 4. The reaction with NEM which alkylates one SH group in the reduced hinge, converting Fc(-SH)$_4$ to Fc(-SH)$_3$, is crucial: in this way we gain an unpaired SH group which can be manipulated without danger of reincorporation into an inter-γ SS bond. After reducing Fcγ1 with DTT at pH 8.0, the resulting Fc(-SH)$_4$ reacts with NEM at pH 5.0, at a molar ratio yielding an average of 0.6 alkylated group per Fc. The reaction is stochastic, and the two important products – capable in principle of exhibiting an active maleimide group after the eventual reaction with the bismaleimide linker – are those molecules with one alkylated group (37%) and those with three (1%). The presence of three alkylated groups means that the hinge cannot be re-closed, and the total extent of alkylation is kept low in order to keep the proportion of such molecules acceptably low (an average of 1.0 alkylated group per Fc yields 42% of molecules with one alkylated group,

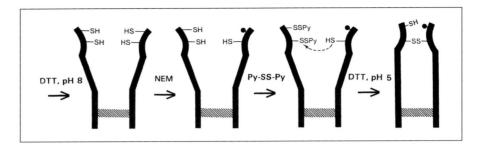

Fig. 4. Reactions leading to Fc-SH, with a closed hinge, from human Fcγ1. After an initial reduction yielding Fc(-SH)$_4$, a limited reaction with NEM alkylates 0.6 SH group per molecule. A random distribution of the alkylated group means that 37% of the molecules are alkylated at one site as depicted. Reaction with Py-SS-Py is shown to close the hinge, and form Fc-SS-Py with the unpaired SH group. Finally, as seen with Fab'-SS-Py, DTT at pH 5 breaks the pyridyl-linked SS bond to produce Fc-SH.

but 5% with three). Molecules alkylated at zero, two or four sites are not expected to exhibit a maleimide group after reacting with the linker, and act simply as carrier. This strategy leads to a low recovery of Fc in the final conjugate, but this is quite acceptable given the modest cost of human IgG.

After alkylation with NEM the Fc(-SH)3.4 reacts with Py-SS-Py at pH 5.0, as described for Fab'-SS-Py. Where homologous cysteine residues are present in SH form the inter-γ SS bond is reconstituted by intramolecular SS interchange, while unpaired SH groups are converted to SS-Py.

Fc-SH and Fc-Maleimide

When the Fc module is required for conjugation, preparation of Fc-maleimide proceeds as described earlier for Fab'-maleimide, via the intermediate form Fc-SH (fig. 4). The yield of maleimide groups after reaction with the PDM linker may be assayed by reaction with, and back titration of, a standard thiol solution. This has averaged about 0.33 per molecule.

Antibody Constructs

Planning the Constructs

A wide range of mosaics may be constructed using the Fab' and Fc modules described above. Other types of modules may of course be added, such as ribosome-inactivating toxins of plant origin [19]. The bonds between the modules will in general be tandem thioether arising from the PDM linker, but where cleavage of the union in vivo is required, as in the toxin-antibody bond of immunotoxins, an SS bond may be used [20].

Is the construct to be univalent or bivalent? With strongly modulating cell-surface molecules, such as IgM targetted by anti-idiotype antibodies, univalency will minimize (although not prevent) modulation in vivo [21]. With some other surface molecules (for example CD20 [22]), modulation does not appear to be a problem, and one may then take advantage of the stronger binding displayed by bivalent antibody. The presence of two Fab' arms also presents the possibility of designing bispecific constructs which target two different molecules on the target cell, thereby increasing the potential density of antibody on the surface, and – if strong binding requires engagement of both arms – providing a further discriminatory function in favor of cells displaying both molecules.

Should the construct contain more than one Fc module? The presence of more than one Fc can enhance both complement activation and antibody-dependent cellular cytotoxicity (ADCC) [23–25], but one should ensure that there is no anticomplementary activity – that is, no spontaneous activation of complement in the absence of antigen [25].

Should the Fc modules have open or closed hinges? IgG molecules with open hinges lack the ability to activate complement, but have not been reported to lose ADCC function [4]. Although one might normally prefer Fc with full effector functions, there may be ·circumstances in which the activation of complement with release of anaphylatoxins leads to undue toxicity and is better avoided.

A Sample Construct: Bispecific Fab₂Fc₂ (Anti-CD20/Anti-CD37)

The preparation of a derivative similar to this, and some of its properties, have been described previously [23]. The modular outline of the present derivative is depicted in figure 5. The specificities of the two arms are directed against differentiation antigens on B lymphocytes, and the derivative is designed to attack B-cell lymphomas displaying these antigens. In invoking either complement lysis or ADCC, it has proved notably more powerful than its univalent equivalents (FabFc anti-CD20, FabFc anti-CD37), used either singly or in combination.

Equal amounts of each Fab'γ partner, stored in the form Fab-SS-Py, are used. One is converted to Fab'-SH by reaction with DTT, separated on Sephadex G-25, and kept cold at pH 5.0 pending conjugation. The second partner is converted sequentially to Fab'SH and Fab'-maleimide, and then added to the Fab'-SH. With each partner present at about 1.3 mg/ml, the coupling reaction is allowed to proceed at 25 °C for 18 h. The protein is then concentrated some 10-fold by a cycle of adsorption and elution on SP

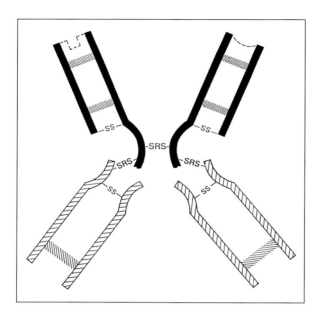

Fig. 5. Topography of bispecific chimeric Fab$_2$Fc$_2$ prepared from murine Fab'γ (solid) and human Fcγ1 (hatched). The figure is idealized in showing one Fc attached to each Fab (some constructs could have both Fc on one Fab'), and in showing each γ-light SS bond reconstituted by the final reaction with cystamine (cysteines from these bonds could in some cases be involved in intermodular links).

Sepharose, and the eluate is led onto Superdex 200 for size-separation of the bispecific dimer. In theory, the dimer now possesses no free SH groups (in actuality <0.3/100 kD), but on reduction will yield eight such groups while its tandem thioether link remains intact. The reduction can conveniently be combined with the elution and gel chromatography just described: elute the SP Sepharose with 2 mM DTT in 0.5 M NaCl, 0.5 M Tris, pH 8.5, and lead the protein into the Superdex column which has been equilibrated with 0.1 M acetate of pH 5.0, but has 0.2 times its volume of the DTT/eluting buffer introduced immediately before the protein. In this way the F(ab')$_2$ is separated from residual monomer and other products while in the presence of DTT, but the last part of the column passage sees it delivered as F(ab')$_2$(-SH)$_8$ in the acetate buffer at pH 5.0.

Add immediately Fc-maleimide in acetate buffer of pH 5.0, and incubate under N$_2$ at 25 °C for 18 h: to obtain Fab$_2$Fc$_2$ as the predominant product the total amount of Fc (two parts carrier, one part reactive protein) should be three times the amount of F(ab')$_2$. The reaction may be ended by alkylating surplus SH groups, or by carrying out an SS-interchange (with 10 mM cysta-mine at pH 8.0) which will reconstitute many of the intramolecular SS bonds in the Fab' modules. The bispecific chimeric construct Fab_2Fc_2 is finally purified by gel chromatography.

References

1   Stevenson GT, Glennie MJ, Kan KS: Chemically engineered chimaeric and multi-Fab antibodies, in Clark M (ed): Protein Engineering of Antibody Molecules for Prophylactic and Therapeutic Applications in Man. Nottingham, Academic Titles, 1993, pp 127–141.
2   King DJ, Turner A, Farnsworth APH, Adair JR, Owens RJ, Pedley RB, Baldock D, Proudfoot KA, Lawson ADG, Beeley NRA, Millar K, Millican TA, Boyce BA, Antoniw P, Mountain A, Begent RHJ, Shochat D, Yarranton GT: Improved tumor targeting with chemically cross-linked recombinant antibody fragments. Cancer Res 1994;54:6176–6185.
3   Seegan GW, Smith CA, Schumaker VN: Changes in quaternary structure of IgG upon reduction of the interheavy-chain disulfide bond. Proc Natl Acad Sci USA 1979;76:907–911.
4   Isenman DE, Dorrington KJ, Painter RH: The structure and function of immunoglobulin domains. II. The importance of interchain disulfide bonds and the possible role of molecular flexibility in the interaction between immunoglobulin G and complement. J Immunol 1975;114:1726–1729.
5   Fujio H, Karush F: Antibody affinity. II. Effect of immunization interval on antihapten antibody in the rabbit. Biochemistry 1966;5:1856–1863.
6   Ashman RF, Metzger H: A Waldenström macroglobulin which binds nitrophenyl ligands. J Biol Chem 1969;244:3405–3411.
7   Padlan EA: Anatomy of the antibody molecule. Mol Immunol 1994;31:169–217.
8   Parham P: On the fragmentation of monoclonal IgG1, IgG2a and IgG2b from BALB/c mice. J Immunol 1983;131:2895–2902.
9   Yamaguchi Y, Kim H, Kato K, Masuda K, Shimada I, Arata Y:Proteolytic fragmentation with high specificity of mouse immunoglobulin G. Mapping of proteolytic cleavage sites in the hinge region. J Immunol Methods 1995;181:259–267.
10 Liu TY: The role of sulfur in proteins; in Neurath H (ed): The Proteins. New York, Academic Press, 1977, pp 239–402.
11 Cecil R, McPhee JR: The sulfur chemistry of proteins. Adv Protein Chem 1959;14:255–389.
12 Cleland WW: Dithiothreitol, a new protective reagent for SH groups. Biochemistry 1964;4:480–482.
13 Grassetti, DR, Murray Jr JF: Determination of sulfhydryl groups with 2,2'- or 4,4'-dithiopyridine. Arch Biochem Biophys 1967;119:41–49.
14 Brocklehurst K: Specific covalent modification of thiols: Applications in the study of enzymes and other biomolecules. Int J Biochem 1979;10:259–274.
15 Grimshaw CE, Whistler RL, Cleland WW: Ring opening and closing rates for thiosugars. J Am Chem Soc 1979;101:1521–1532.
16 Hamblin TJ, Cattan AR, Glennie MJ, MacKenzie MR, Stevenson FK, Watts HF, Stevenson GT: Initial experience in treating human lymphoma with a chimeric univalent derivative of monoclonal anti-idiotype antibody. Blood 1987;69:790–797.
17 Glennie MJ, McBride HM, Worth AT, Stevenson GT: Preparation and performance of bispecific F(ab'γ)₂ antibody containing thioether-linked Fab'γ fragments. J Immunol 1987;139:2367–2375.
18 Schott ME, Frazier KA, Pollock DK, Verbanac KM: Preparation, characterization, and in vivo biodistribution properties of synthetically cross-linked multivalent antitumor antibody fragments. Bioconjugate Chem 1993;4:153–165.
19 Vitetta ES, Thorpe PE, Uhr JW: Immunotoxins – magic bullets or misguided missiles. Trends Pharmacol Sci 1993;14:148–154.
20 Wawrzynczak EJ, Cumber AJ, Henry RV, Parnell GD, Westwood JH: Structural factors influencing the pharmacokinetics and stability of immunotoxins. Biochem Soc Trans 1992;20:738–743.

21 Lane AC, Foroozan S, Glennie MJ, Kowalski-Saunders P, Stevenson GT: Enhanced modulation of antibodies coating guinea pig leukemic cells in vitro and in vivo. The role of $FC\gamma R$ expressing cells. J Immunol 1991;146:2461–2468.

22 Press OW, Appelbaum F, Ledbetter JA, Martin PJ, Zarling J, Kidd P, Donnall Thomas E: Monoclonal antibody 1F5 (anti-CD20) serotherapy of human B cell lymphomas. Blood 1987;69:584–591.

23 Stevenson GT, Pindar A, Slade AJ: A chimeric antibody with dual Fc regions (bisFabFc) prepared by manipulations at the IgG hinge. Anti-Cancer Drug Design 1989;3:219–230.

24 Caron PC, Laird W, Sung Co M, Avdalovic NM, Queen C, Scheinberg DA: Engineered humanized dimeric forms of IgG are more effective antibodies. J Exp Med 1992;176:1191–1195.

25 Smith RIF, Morrison SL: Recombinant polymeric IgG: An approach to engineering more potent antibodies. Biotechniques 1994;12:683–688.

George T. Stevenson, Tenovus Research Laboratory, Southampton University Hospital, Southampton SO16 6YD (UK)

Capra JD (ed): Antibody Engineering.
Chem Immunol. Basel, Karger, 1997, vol 65, pp 73–87

....

Structure and Function in IgA

J. Mark Hexham, Leonidas Carayannopoulos, J. Donald Capra

Molecular Immunology Center, Department of Microbiology, UT Southwestern
Medical Center at Dallas, Dallas, Tex., USA

Immunoglobulins mediate humoral immunity by attaching to foreign antigens and then recruiting effector modalities (e.g. complement, granulocytes, cytotoxic T-cells) to destroy and clear the antigens. Each of the five immunoglobulin isotypes interacts with specific effector systems via its constant region Fc domain. Due to the immunological and clinical relevance of immunoglobulin-effector interactions, antibody structure/function is the subject of intense and fruitful research [reviewed in ref. 1].

The mucosal surfaces of the body represent the largest area of exposure of the body to external pathogens, 400 m² compared to only 1.8 m² of skin area [2]. IgA is the immunoglobulin subclass primarily responsible for humoral immune protection at this large exposed surface. IgA has several unique properties which enable it to carry out its specialized function in mucosal defense. Firstly, IgA can intracellularly associate with J chain via a cysteine in the C-terminal 'tail' to form dimeric IgA (dIgA) [3] which can be bound by the polymeric immunoglobulin receptor (pIgR) and transported across mucosal epithelia [4, 5] to be released as secretory IgA (sIgA). The IgA thus transported serves as the first line of humoral defense at mucosal surfaces [6]. Consequently, it is also the target of specific receptors and proteases produced by a number of pathogenic bacteria in an effort to evade IgA-mediated immunity [7]. Blood also contains a large quantity (average 2 mg/ml [8]) of predominantly monomeric IgA; this circulating pool is largely independent of the mucosal pool in humans [9]. Secondly, IgA is known to bind to a recently cloned [10–12] Fcα receptor (mFcαR, CD89) on the surface of eosinophils [13], neutrophils [14], and monocyte/macrophages [10], thus triggering effector responses. Thirdly, B lymphocytes and T lymphocytes possess surface receptors for Fcα through which immunoregulatory signals are thought to be transmitted [15]. Fourthly, a receptor specific for sIgA has been identified on eosinophils which may have a role in eosinophil activation and degranulation during the inflammatory response [16]. Lastly, IgA is thought to activate complement through the alternative pathway [17, 18]. Due to these functional properties, especially the ability to be transported through cells, IgA not only plays a role in host defense against viruses and bacteria, but is also potentially critical for neutralization of intracellular viruses in tissues expressing pIgR [19], and destruction of helminths, protozoans and other eukaryotic parasites [20]. Several immuno-logic disease processes also are mediated by IgA, including IgA glomerulo-nephritis [21] and possibly the exacerbation of allergic asthma [22, 23]. These properties of IgA depend on the ability of effector molecules such as the pIgR, mFcαR or C3 to recognize specific sites on the surface of the Fcα protein. Further elucidation of these sites should provide new insights into the physio-logical role of IgA.

Here we discuss the structure of IgA, its interaction with other molecules and the functional aspects of these structures. We concentrate on the human IgA system apart from a brief phylogenetic comparison and when certain systems have been initially better characterized, eg. rabbit pIgR [24].

Biochemistry of IgA

Structure of IgA

IgA is a tetrameric protein comprising two identical light chains (κ or λ) and two identical heavy chains (α) which endow IgA with its biologically specific properties (fig 1). In the human, there are two IgA isotypes, IgA1 and IgA2, present as a result of the duplication of a large segment of the human heavy chain locus [26]. The overall domain structure of IgA appears to resemble IgG in that it contains three constant domains (Cα1–Cα3), with a hinge region between the Cα1 and Cα2 domains. IgA1 and IgA2 differ largely in the hinge region where IgA2 lacks a proline-rich sequence of 13 amino acids, present in IgA1 [26-28]. These differences in hinge regions confer differential suscepti-bility to bacterial proteases on the different IgA molecules (see below). All IgA isotypes (as well as IgMs) have an additional 18 amino acid C-terminal to the Cα3 domain which enables polymeric immunoglobulin formation [29, 30].

IgA2 has three allotypes IgA2m(1), IgA2m(2) and a third allotype of IgA2, recently characterized at the genetic level, which will likely be named IgA2m(3) [28, 31, 32, respectively]. The IgA2m(1) allotype appears to be a hybrid gene possibly arising from a recombination or gene conversion involving IgA1 and IgA2m(2), thus the Cα3 domain of IgAm(1) is identical to that of IgA1, whereas the Cα1 Cα2 domains are clearly related to IgA2m(2) [28]. The

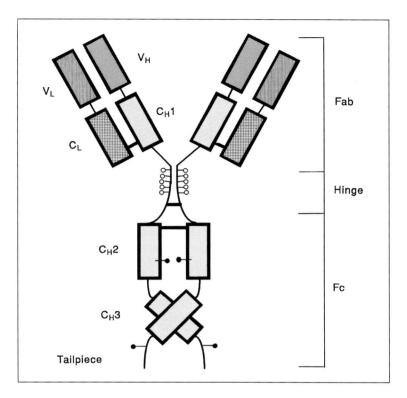

Fig. 1. Human IgA1 monomer. The N-linked and O-linked glycosylation sites are indicated by filled and open circles, respectively. Disulfide bonds are indicated by dark bars.

novel 'IgA2m(3)' molecule is also a hybrid molecule and comprises a $C\alpha1$ sequence identical to IgA2m(2), a $C\alpha2$ domain identical to IgA2m(1) or IgAm(2) and a $C\alpha3$ sequence identical to IgA2m(1) or IgA1 [32]. In IgA1, the light chains are thought to disulfide-bond to the heavy chains via Cys133 in the $C\alpha1$ domain (fig 1). In IgA2m(1), however, the light and heavy chains are not disulfide-linked [33], this is due to the presence of Asp at this position. In contrast IgA2m(2) and 'IgA2m(3)' form heavy chain-light chain disulfide bonds presumably to another heavy chain cysteine residue.

In general, IgA and IgM are more heavily glycosylated (7–10%) than IgG (3%) [34], suggesting a role for the glycosylation in protection of these secreted Igs in the mucosal milieu. Within the IgA isotypes, IgA2 is more highly glycosylated than IgA1 [35]. The IgA1 heavy chain contains two potential N-linked glycosylation sites at Asn263 within the $C\alpha2$ domain and at Asn459 within the tailpiece [36]. IgA2m(1) contains two additional sites at positions Asn166 in the $C\alpha1$ domain and Asn337 in the $C\alpha2$ domain. IgA2m(2) contains

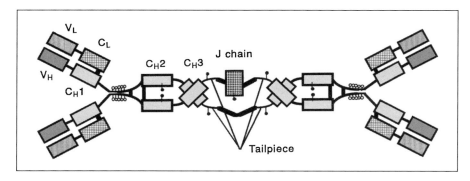

Fig. 2. Assembly of dimeric IgA1 with J chain. The positions of the J chain-tailpiece and tailpiece-tailpiece disulfide bonds are indicated by dark bars and are as determined by Bastian et al. [40].

all these sites plus one more at position Asn211 in the $C\alpha2$ domain. IgA1 alone possesses the O-linked glycosylation sites at serines 224, 230, 232, 238 and 240 [37]. These residues are not present in all IgA2 allotypes due to either the deletion of the hinge region (224, 230, 232) or their substitution by proline (238, 240), thus IgA2 has no O-linked sugar residues. The O-linked glycosylation of IgA1 facilitates affinity purification of IgA1 using the lectin jacalin as a ligand [38].

Polymerization of IgA

Serum IgA is present in monomeric or polymeric forms, which are mostly dimers though tetramers and higher polymers have been observed. This polymerization process is unique to IgA and IgM and is mediated by the 18 amino-acid tailpiece in conjunction with the J chain, one molecule of which is present per dimer of IgA [39]. The J chain is a small glycoprotein (15 kD) component of polymeric IgA and IgM [3]. J chain contains six interchain disulfide bonds and two additional cysteine residues at positions 15 and 79 which form disulfide bonds to the penultimate cysteine of one heavy chain in each IgA monomer subunit [40]. Thus J chain bridges the two IgA monomers, while the other two tailpiece cysteine residues bind directly, stabilizing the dimeric molecule further (fig. 2).

While the presence of the tailpiece, in particular the penultimate cysteine residue, has been shown to be necessary for dimerization of IgA [41] or pentamerization of IgM [30], the precise role of the J chain is unclear. IgM can form tetramers and hexamers in the absence of J chain [42, 43], and seems to have a more flexible or redundant mechanism for disulfide bond-mediated polymerization than does IgA [for a review, see ref. 44]. In the absence of either J chain or the penultimate IgM cysteine (residue 575, polymerization appears to occur via an additional cysteine (residue 414), present in the Cμ3 domain [43]. IgA however does not possess a cysteine residue at the analogous position within the Cα2 domain and this presumably is one factor in the difference in polymerization observed between IgA and IgM. In spite of this, there is evidence of IgA polymerization in the absence of J chain from J-chain-knockout mice [45] albeit at a much lower level. In addition, IgA polymers with less than one J chain molecule per polymer have been observed in human myelomas [46–48]. By contrast, a transgenic plant system in which IgA was expressed in conjunction with J chain and secretory component required J chain coexpression to generate dimeric IgA [49]. While these situations are all different and prone to various artifacts it is clear that optimal IgA polymerization occurs in the presence of J chain.

Fusion of the IgM tailpiece to the C-terminus of human IgG results in polymerization of all four IgG subclasses [50]. Recent experiments comparing the ability of the IgA and IgM tailpieces and various α/μ chimeric tailpieces to polymerize antibodies demonstrate the functional similarities between these two homologous 18 amino-acid sequences [51]. The IgA and IgM tailpieces are identical at 11 out of 18 residues and substitution of IgA for IgM tailpiece had no effect on the polymerization of IgM with the usual pattern of pentamers and hexamers being observed. One of the hybrid α/μ tailpieces caused an intermediate degree of polymerization with more monomers, dimers and tetramers being formed. The addition of either the IgM or IgA tailpiece to IgG caused polymerization of IgG with all polymeric forms from dimers to pentamers being present. These results indicate that while the tailpiece mediates polymerization, there are other determinants on the heavy chain molecule which control the level of polymerization. We have also observed polymerization of human IgG1 by fusing the IgA tailpiece onto the molecule.

The role of the J chain is certainly to aid polymerization of IgM and IgA and to stabilize the polymeric molecules in the 'external' or mucosal environment. Presumably, the differences observed in the behavior of these two isotypes with respect to polymerization are due to the difference in their structures and the ways these interact with the J chain. The J chain may have a role in the interaction of polymeric immunoglobulins with the pIgR [47, 52], and the J-chain-knockout mice have greatly decreased secretory IgA levels [45].

Secretory IgA

Secretory IgA plays a major role in humoral defense at mucosal surfaces, which constitute the largest area of the body normally exposed to external pathogens [2]. Polymeric immunoglobulins (IgA and IgM) are transported to the mucosal surface via a highly specific receptor-mediated process [4, 5].

Dimeric IgA is specifically bound by the pIgR, expressed on the basolateral surface of mucosal epithelial cells, and transported through these cells to be secreted at the mucosal surface. Secretory IgA (sIgA) retains most of the extracellar region of the pIgR, termed secretory component (SC), covalently bound to one of the IgA monomers [53].

The pIgR is a transmembrane glycoprotein with five immunoglobulin superfamily homology domains (I–V) in the extracellular region [4, 54]. The four most N-terminal domains share more sequence homology with immuno-globulin V region genes while the fifth domain is more like an immunoglobulin C region. To date, pIgR genes from five mammalian species have been sequenced and cloned. These are the rabbit [4], human [54, 55], rat [56], mouse [57] and cow [58]. The five molecules share between 44 and 86% homology, with mouse and rat being most closely related; human, cow and rabbit sequences are increasingly divergent from the mouse sequence. The human sequence contains 20 cysteine residues which form disulfide bridges to stabilize the immunoglob-ulin-like domains, the number of these varies slightly between species but most of them [17] are perfectly conserved in all species studied. The N-linked glycosylation pattern is more variable across the species with human and mouse containing seven potential glycosylation sites compared to four in the rat, three in the cow and two in the rabbit. In addition, the positions of these glycosylation sites varies, and none of them is completely conserved in any of the five organisms.

The complete genomic organization of the human pIgR gene has been determined [55] and some information is available for the rabbit [59]. The human gene is encoded by 11 exons and the immunoglobulin-like extracellular domains I, IV and V are each encoded by a separate exon. However domains II and III are encoded together on a single larger exon. Lower-molecular-weight forms of the pIgR have been observed in the rabbit at both the RNA [59] and protein level [60] and in the cow at the RNA level [61]. These lower-molecular-weight forms are the product of an alternative splice variant of the pIgR, lacking domains II and III. However, there is no evidence that the human, rat and mouse genes give rise to anything but a single message. In the rabbit system this lower-molecular-weight form is still active in both IgA binding and transcytosis [62].

The primary site of interaction of pIgR with dIgA is in domain I [63], which participates in a high affinity (10^8 M^{-1}), non-covalent interaction [64]. Further mapping of the dimeric IgA binding site within domain I of the pIgR has identi-fied a peptide comprising residues 15–37 of human pIgR which binds dIgA [65]. This peptide showed high interspecies conservation and was also identified by a monoclonal antibody which was able to inhibit binding of dIgA to pIgR [65, 66]. In these studies the affinity of the interaction of the peptide with dIgA was reduced about 100-fold compared to the intact molecule. Similarly, the specificity of the interaction was reduced such that binding of monomeric IgG to the peptide was now detectable though at lower affinity than dIgA and pIgM binding. These observations demonstrate that while this peptide may constitute a central part of the binding site there are other regions of domain I that make important contributions to dIgA binding. These authors also showed that a deglycosylated form of the pIgR molecule was still able to bind dIgA.

A mutational approach, based on modeling of the domain I sequence on known immunoglobulin variable domain structures, demonstrated that the loops in analogous positions to the three V region CDRs made up the dIgA binding site [67]. The peptide identified by Bakos et al. [65, 66] overlaps with part of the CDR1-like loop identified in this study, namely residues 31–40. The role of the CDR2-like (residues 56–60) and CDR3-like (residues 100–105) loops appears to be equally important as mutations in both of these regions abrogated binding. This suggests that the interaction of the pIgR with dIgA is similar to the interaction of antibody with antigen, or to be more precise, the interaction of a single V domain with antigen.

The association of IgA with the pIgR appears to be a two-stage process [8], the first being the binding of dIgA to domain I as discussed above. The second stage of the interaction between pIgR and dIgA involves covalent binding of domain V to the Fc of one of the subunits in dIgA [68, 69]. This single disulfide is formed between Cys467 in domain V of secretory component and Cys311 located in the Cα2 domain of a heavy chain in one IgA subunit [70]. This disulfide formation appears to be a late event in the secretion pathway and is not absolutely necessary for transcytosis [71].

Effector Functions of IgA

Protective antibodies of the IgA isotype have been documented against a wide range of human pathogens including viruses such as HIV [72] and influenza A [73], bacteria [74, 75] bacterial toxins and macroscopic parasites [76]. There are several mechanisms by which IgA exerts its antimicrobial effect and they may be divided into active and passive mechanisms. The passive mechanisms are independent of the effector arm of the immune system and include neutralization of viruses by blocking of their receptors for host cell protein and inhibition of bacterial motion by aggregation.

However, in addition to these mechanisms there are a number of active processes unique to the IgA isotype; firstly, IgA binds to its cognate Fc receptor termed FcαR or CD89 [11]. This receptor is expressed on neutrophils, eosinophils and cells of the monocyte/macrophage series and binds IgA/antigen complexes, activates the cells, induces phagocytosis and can thus lead to the elimination of pathogenic organisms. The human myeloid FcαR gene encodes a transmembrane glycoprotein of 30 kD with two immunoglobulin-like domains in the extracellular region and six potential N-linked glycosylation sites [12]. The CD89 antigen exhibits differential glycosylation leading to a wide molecular weight range of 55–100 kD in various cell types. We have recently mapped the FcαR binding site on human IgA1 to a region comprising the boundary between the C_H2 and C_H3 domains both of which appear to contribute to the binding site [77]. The position of this site is hinge distal whereas the FcR sites for IgG and IgE are in hinge proximal regions of these molecules [78, 79]. This C_H2/C_H3 boundary site seems well suited to a role in biological recognition as it is also used by *staphylococcal* protein A to bind IgG [80], the rat Fcn receptor to recognize IgG [81] and an as yet unidentified receptor system involved in IgG catabolism [82].

IgA is unable to activate complement via the classical pathway, however complement activation via the alternative pathway has been demonstrated in chimeric antibodies with human [17, 18], rabbit [83] and rat [84] Fc regions. On the other hand some studies have demonstrated little or no activation of the alternative pathway by IgA and argue that IgA has a noninflammatory role in immune regulation [8, 85].

The above mechanisms are shared by antibodies of the IgG isotype; however, there are two unique ways in which the secretory nature of IgA can lead to elimination of pathogens. The pIgR pathway can lead to elimination of immune complexes from the submucosal area [86, 87]. Secondly, the transcytosis pathway enables IgA to act intracellularly to neutralize viruses, at least in the transcytotic compartments [19].

IgA Comparative Biology

IgA has been detected in all mammals studied [88] as well as marsupials [89], together with IgM and the other components of the secretory immune system.

Avian species also express IgA and possess a secretory immune system, with chicken being the best characterized example [90]. Many of these comparative biology studies were based on serum cross-reactivity but a cDNA encoding the chicken IgA gene has been cloned and shares 37% identity with murine IgA [91]. The sequence identities between chicken IgA and the IgM, IgG and IgD are 29, 30 and 26%, respectively. Interestingly the chicken IgA contains four constant domains, like IgM, suggesting that IgA originally arose from IgM and that the C_H2 domain of an ancestral IgA gene was lost at some point during the evolution of mammals.

In studies of reptiles, IgM but not IgA has been detected [92], similar observations were made in fish [93]. Both of these groups appear to possess a secretory immune system, based on IgM production.

The development of secretory immunity based on IgM-like molecules seems to have occurred early in evolution. The divergence of IgA from an avian IgM-like precursor has occurred at a later time. This order of appearance in the phylogenetic classification is recapitulated during mammalian development with IgA being the last immunoglobulin isotype to be expressed at adult levels in neonates.

The lagomorphs, exemplified by rabbit, are unique among mammalian species in having 13 genes in their Cα locus [94], whereas other species have one or two Cα genes. All 13 rabbit Cα genes appeared to be expressible as proteins and 12 were produced in vitro all of these were transcytosed by the pIgR [95]. Human IgA1 and IgA2 are differentially expressed at mucosal sites. Similarly, expression of 10 of the 13 rabbit IgA genes was observed in vivo, with expression levels varying greatly between different locations [96]. The extent of the lagomorph IgA gene family is probably related to their unique digestive system, though it is unclear whether there are any lagomorph IgA specific bacterial proteases which may have driven this diversification, as appears to be the case in humans [97].

J-chain expression has been detected in many mammalian species and the cDNA sequences of the human [98], mouse [99], cow [100] as well as the protein sequences of the rabbit [101] and bullfrog [102] J chain are available. Identifiable J chain has an evolution which antedates the appearance of immunoglobulins as it is found in 14 non-immunoglobulin-expressing invertebrates such as the earthworm and slug [103]. The J chain was detected in the mucus cells, the intestinal epithelial cells and the macrophage-like cells. The invertebrate J chain clearly has a role different from antibody polymerization, but its localization may indicate a general mucosal defense function of some kind. The high level of conservation observed between vertebrate and non-vertebrate sequences suggests that the invertebrate functions of the J chain may still be preserved in vertebrates.

Molecular Analysis of IgA

For mutational analysis of antibody structure/function relationships it is essential to have access to good expression systems which permit facile manipulation of the molecules. Human IgA has been expressed in a range of systems which have various advantages and disadvantages.

IgA1 and 2 have been expressed in chinese hamster ovary and COS cells; the protein produced was authentic by a number of criteria [104]. IgA1 has been expressed in insect cells infected with baculoviruses [18], this material could be produced as dimers with J-chain, bound the FcαR and C3 and was glycosylated on both the N-linked and O-linked sites. One disadvantage of the insect cell expression system is that the glycosylation is of the high mannose type and not the complex carbohydrate found in mammalian proteins. The insect cell system has also been used to express human secretory component with dimeric IgA binding activity, the same study also expressed SC in a vaccinia virus system [105].

Transgenic plants have also been used to good effect and this system was able to produce not only dimeric IgA including J chain but also secretory IgA with SC already bound to the dimeric IgA [106]. This system is ideal for the large-scale production of assembled dimeric and secretory IgA for therapy or structural studies; however, it does not easily permit manipulation via mutation of the expressed molecules; furthermore the plant cell glycosylation patterns will be different. Transgenic mice, pigs and sheep expressing IgA have also been created [107]. Apart from the ability to make large amounts of protein with correct mammalian glycosylation, the latter experiments offer the potential benefit of 'pre-programmed' germline-encoded immunity in large domesticated animals.

Future Directions

The most important future developments for the use of IgA in therapy are in the field of mucosal vaccination. As the site of entry of most pathogens, the mucosal surface is the ideal site to induce immunity to prevent entry of such organisms. Such vaccination programs are already under evaluation and may hopefully present solutions to HIV transmission, infectious respiratory diseases and other diseases with gastrointestinal transmission [108, 109].

The technique of using intracellular antibodies to fight disease or inhibit intracellular mechanisms has recently been developed [110]. This method uses gene therapy approach where the antibody genes are introduced into the cell and has the strength that expression of the antibody can be directed to the appropriate subcellular compartment. The observation that IgA can also neutralize intracellular viruses [19] in epithelial cells during transcytosis, makes it a possible candidate for use in this kind of therapy. Clearly the limitation is that only pIgR-expressing cell types can be targeted but it may still be worth pursuing for epithelial cell specific viruses and tumors.

The pIgR system may also be exploited as a specific drug or gene delivery system to target mucosal areas. Recently gene therapy of human respiratory epithelial cells exploiting the pIgR pathway has been demonstrated, at least in an ex vivo system [111]. This study actually used an anti-pIgR Fab cross-linked to polylysine to introduce a reporter gene containing plasmid into the epithelial cells. In the future, engineered IgA molecules may allow direct targeting of viral vectors or liposomes to mucosal epithelial cells for gene therapy.

The wealth of information that is becoming available on antibody effector functions may, in the future, permit construction of chimeric antibodies with tailor-made effector functions purpose-built for therapeutic applications. This type of approach, coupled with the enormous advances made in engineering and enhancing antibody specificities [112, 113] may enable the production of a new generation of antibodies where both the antigen specificity and effector functions can be brought together in powerful new therapeutic antibodies.

References

1   Burton DR, Woof JM: Human antibody effector functions. Adv Immunol 1992;51:1–84.
2   Childers NK, Bruce MG, McGhee JR: Molecular mechanisms of immunoglobulin A defense. Annu Rev Microbiol 1989;43:503–536.
3   Koshland ME: The coming of age of the immunoglobulin J chain. Annu Rev Immunol 1985;3: 425–453.
4   Mostov KE, Blobel G: A transmembrane precursor of secretory component. The receptor for transcellular transport of polymeric immunoglobulins. J Biol Chem 1982;257:11816–11821.
5   Mostov KE: Transepithelial transport of immunoglobulins. Ann Rev Immunol 1994;12:63–84.
6   Underdown BJ, Schiff JM: Immunoglobulin A: Strategic defense initiative at the mucosal surface. Ann Rev Immunol 1986;4:389–417.
7   Kilian M, Mestecky J, Russell MW: Defense mechanisms involving Fc-dependent functions of immunoglobulin A and their subversion by bacterial immunoglobulin A proteases. Microbiol Rev 1988;52:296–303.
8   Mestecky J, McGhee JR: Immunoglobulin A (IgA): Molecular and cellular interactions involved in IgA biosynthesis and immune response. Adv Immunol 1987;40:153–245.
9   Jonard PP, Rambaud JC, Dive C, Vaerman JP, Galian A, Delacroix D: Secretion of immunoglobulins and plasma proteins from the jejunal mucosa. Transport rate and origin of polymeric immunoglobulin A. J Clin Invest 1984;74:525–535.
10 Shen L, Lasser R, Fanger MW: My 43, a monoclonal antibody that reacts with human myeloid cells inhibits monocyte IgA binding and triggers function. J Immunol 1989;143:4117–4122.
11 Monteiro RC, Kubagawa H, Cooper MD: Cellular distribution, regulation and biochemical nature of an Fcα receptor in humans. J Exp Med 1990;148:597–613.
12 Maliezewski CR, March CJ, Schoenborn MA, Gimpel S, Shen L: Expression cloning of a human Fc receptor for IgA. J Exp Med 1990;172:1665–1672.
13 Monteiro RC, Hostoffer RW, Cooper MD, Bonner JR, Gartland GL, Kubagawa H: Definition of immunoglobulin A receptors on eosinophils and their enhanced expression in allergic individuals. J Clin Invest 1993;92:1681–1685.
14 Kurita T, Kiyono H, Komiyama K, Grossi CE, Mestecky J, McGhee JR: Isotype-specific immunoregulation; characterization and function of Fc receptors on T-T hybridomas which produce murine IgA-binding factor. J Immunol 1986;136:3953–3960.
15 Weisbart RH, Kacena A, Schuh A, Golde DW: GM-CSF induces human neutrophil IgA-mediated phagocytosis by an IgA Fc receptor activation mechanism. Nature 1988;332:647–648.
16 Lamkhioued B, Gounni AS, Gruart V, Pierce A, Capron A, Capron M: Human eosinophils express a receptor for secretory component. Role in secretory IgA-dependent activation. Eur J Immunol 1995;25:117–125.
17 Lucisano-Valim YM, Lachmann PJ: The effect of antibody isotype and antigenic epitope density on the complement-fixing activity of immune complexes: A systematic study using chimaeric anti-NIP antibodies with human Fc regions. Clin Exp Immunol 1991;84:1–8.
18 Carayannopoulos LN, Max EE, Capra JD: Recombinant human IgA expressed in insect cells. Proc Natl Acad Sci USA 1994;91:8348–8352.

19   Mazanec MB, Kaetzel CS, Lamm ME, Fletcher D, Nedrud JG: Intracellular neutralization of virus by IgA antibodies. Proc Natl Acad Sci USA 1992;89:7252–7256.

20   Grzych JM, Grezel D, Xu C, Neyrinck JL, Capron M, Ouma JH, Butterworth AE, Capron A: IgA antibodies to a protective antigen in human *Schistosoma mansoni*. J Immunol 1993;150:527–535.

21   Clarkson AR, Woodroffe AJ, Bannister KM, Lomax-Smith JD, Aarons I: The syndrome of IgA nephropathy. Clin Nephrol 1984;21:7–14.

22   Kitani S, Ito K, Miyamoto T: IgG, IgA, and IgM antibodies to mite in sera and sputa from asthmatic patients. Ann Allergy 1985;55:612–620.

23   Popper H, Pongratz M, Lanzer G: IgA2–alveolitis and eosinophilic pneumonia – a possibly virus-triggered allergy. Allergol Immunopathol (Madr) 1982;10:177–184.

24   Mostov KE, Freidlander M, Blobel G: The receptor for transepithelial transport of IgA and IgM contains multiple immunoglobulin-like domains. Nature 1984;308:3743.

25   Flanagan JG, Rabbitts TH: Arrangement of human immunoglobulin heavy chain gene segments implies evolutionary duplication of a segment containing γ, ε and α genes. Nature 1982;300:709–711.

26   Torano A, Putnam FW: Complete amino acid sequence of the α2 heavy chain of a human IgA2 immunoglobulin of the A2 in (2) allotype. Proc Natl Acad Sci USA 1978;75:966–969.

27   Putnam FW, Liu YSV, Low TLK: Primary structure of a human IgA1 immunoglobulin. J Biol Chem 1979;254:2865–2874.

28   Tsuzukida Y, Wang CC, Putnam FW: Structure of the A2m(1) allotype of human IgA-A recombinant molecule. Proc Natl Acad Sci USA 1979;76:1104–1108.

29   Garcia-Pardo A, Lamm ME, Plaut AG, Frangione B: J chain is covalently bound to both monomer subunits in human secretory IgA. J Biol Chem 1981;256:11734–11738.

30   Davis AC, Roux KH, Schulman MJ: On the structure of polymeric IgM. Eur J Immunol 1988;18: 1001–1008.

31   Torano A, Tsuzukida Y, Liu YSV, Putnam FW: Location and significance of oligosaccharides in human IgA1 and IgA2 immunoglobulins. Proc Natl Acad Sci USA 1977;74:2301–2305.

32   Chintalaruvu KR, Raines M, Morrison SL: Divergence of human α-chain constant region gene sequences: A novel recombinant α2 gene. J Immunol 1994;152:5299-5304.

33   Grey HM, Abel CA, Yount WJ, Kunkel HG: A subclass of human αA globulins (αA2) which lacks the disulfide bonds linking heavy chains and light chains. J Exp Med 1968;128:1223–1236.

34   Carayannopoulos LN, Capra JD: Immunoglobulins: Structure and Function; in Paul WE (ed): Fundamental Immunology. New York, Raven Press, 1993, pp 283–314.

35   Tomana M, Neidermeyeier W, Mestecky J, Skvaril F: The differences in carbohydrate composition between the subclasses of IgA immunoglobulins. Immunochemistry 1976;13:324–328.

36   Baenziger J, Kornfield S: Structure of the carbohydrate units of human IgA1 immunoglobulin. I. Composition, glycopeptide isolation and structure of the asparagine-linked oligosaccharide units. J Biol Chem 1974;249:7260–7269.

37   Baenziger J, Kornfield S: Structure of the carbohydrate units of human IgA1 immunoglobulin. II. Structure of the O-linked oligosaccharide units. J Biol Chem 1974;249:7270–7281.

38   Roque-Barriera MC, Campos-Neto A: Jacalin: An IgA-binding lectin. J Immunol 1985;134:1740–1743.

39   Zikan J, Mestecky J, Kulhavy R, Bennet JC: The stochiometry of J chain in human dimeric IgA. Mol Immunol 1986;23:541–544.

40   Bastian A, Kratzin H, Eckart K, Hilschmann N: Intra- and interchain disulfide bridges of the human J chain in secretory immunoglobulin A. Biol Chem Hoppe-Seyler 1992;373:1255–1263.

41   Despont JPJ, Abel CA, Grey HM, Penn GM: Structural studies on a human IgA1 myeloma protein with a carboxy-terminal deletion. J Immunol 1974;112:1517–1525.

42   Cattaneo A, Neuberger MS: Polymeric immunoglobulin M is secreted by transfectants of non-lymphoid cells in the absence of immunoglobulin J chain. EMBO J 1987;6:2753–2758.

43   Davis AC, Roux KH, Pursey J, Shulman MJ: Intermolecular disulfide bonding in IgM: Effects of replacing cysteine residues in the μ heavy chain. EMBO J 1989;8:2519–2526.

44   Davis AC, Schulman MJ: IgM molecular requirements for its assembly and function. Immunol Today 1989;10:118–126.

45   Hendrickson BA, Conner DA, Ladd DJ, Kendall D, Casanova JE, Corthesy B, Max EE, Neutra MR, Seidman CE, Seidman JG: Altered hepatic transport of immunoglobulin A in mice lacking the J chain. J Exp Med 1995;182:1905–1911.

46   Brandtzaeg P: Complex formation between secretory component and human immunoglobulins related to their content of J chain. Scand J Immunol 1976;5:411–419.

47   Brandtzaeg P, Prydz H: Direct evidence for an integrated function of J chain and secretory component in epithelial transport of immunoglobulins. Nature 1984;311:71–73.

48   Tomasi TB, Czerwinski DS: Naturally occurring polymers of IgA lacking J chain. Scand J Immunol 1976;5:647–653.

49   Ma JKC, Hiatt A, Hein M, Vine ND, Wang F, Stabila P, van Dolleward C, Mostov KE, Lehner T: Generation and assembly of secretory antibodies in plants. Science 1995;268:716–719.

50   Smith RIF, Coloma MJ, Morrison SL: Addition of a μ-tailpiece to IgG results in polymeric antibodies with enhanced effector fuctions including complement-mediated cytolysis by IgG4. J Immunol 1994; 154:2226–2236.

51   Sorensen V, Rasmussen IB, Norderhaug L, Natvig I, Michaelsen TE, Sandlie I: Effect of the IgM and IgA secretory tailpieces on polymerization and secretion of IgM and IgG. J Immunol 1996; 156:2858–2865.

52   Schiff JM, Fisher MM, Underdown BJ: Secretory component as the mucosal transport receptor: Separation of physicochemically analogous human IgA fractions with different receptor-binding capacities. Mol Immunol 1986;23:45–56.

53   Underdown BJ, De Rose J, Plaut A: Disulfide bonding of secretory component to a single monomer subunit in human secretory IgA. J Immunol 1977;118:1816–1821.

54   Eiffert H, Quentin E, Decker J, Hillemeir S, Hufschmidt M, Klingmuller D, Weber MH, Hilschmann N: Die Primastruktur der menschlichen freien Sekretkomponente und die Anordnung der Disulfidbrücken. Hoppe-Seylers Z Physiol Chem 1984;365:1489–1495.

55   Krajci P, Kvale D, Tasken K, Brandzaeg P: Molecular cloning and exon-intron mapping of the gene encoding human transmembrane secretory component (the poly-Ig receptor). Eur J Immunol 1992;22:2309–2315.

56   Banting G, Brake B, Braghetta P, Luzio JP, Stanley KK: Intracellular targetting signals of polymeric immunoglobulin receptors are highly conserved between species. FEBS Lett 1989;254:177–183.

57   Piskurich JF, Blanchard MH, Youngman KR, France JA, Kaetzel CS: Molecular cloning of the mouse polymeric Ig receptor. Functional regions of the molecule are conserved among five mammalian species. J Immunol 1995;154:1735–1747.

58   Verbeet MP, Vermeer H, Warmerdam GCM, de Boer HA, Lee SH: Cloning and characterization of the bovine polymeric immunoglobulin receptor encoding cDNA. Gene 1995;164:329–333.

59   Deitcher DL, Mostov KE: Alternate splicing of rabbit polymeric immunoglobulin receptor. Mol Cell Biol 1986;6:2712–2715.

60   Frutiger S, Hughes J, Fonck C, Jaton JC: High and low molecular weight rabbit secretory components. J Biol Chem 1987;262:1712–1715.

61   Kulseth MA, Krajci P, Myklebost O, Rogne S: Cloning and characterization of two forms of the bovine polymeric immunoglobulin receptor cDNA. DNA Cell Biol 1995;3:251–256.

62   Kuhn L, Kocher HP, Hanly WC, Cook L, Jaton JC: Structural and genetic heterogeneity of the receptor mediating translocation of immunoglobulin A dimer antibodies across epithelia in the rabbit. J Biol Chem 1983;258:6653–6659.

63   Frutiger S, Hughes GJ, Hanly WC, Kingzette M, Jaton JC: The amino terminal domain of rabbit secretory component is responsible for noncovalent binding to immunoglobulin A dimers. J Biol Chem 1986;261:16673–16681.

64   Kuhn LC, Kraehenbuhl JP: Interaction of rabbit secretory component with rabbit IgA dimer. J Biol Chem 1979;254:11066–11071.

65.   Bakos M, Kurowsky A, Goldblum RM: Characterization of a critical binding site for human polymeric Ig on secretory component. J Immunol 1991;147:3419–3426.

66   Bakos M, Kurowsky A, Woodward CS, Denney RM, Goldblum RM: Probing the topography of free and polymeric Ig-bound human secretory component with monoclonal antibodies. J Immunol 1991;146:162–168.

67 Coyne RS, Siebrecht M, Pietsch MC, Casanove JE: Mutational analysis of polymeric immunoglob-ulin receptor/ligand interactions. J Biol Chem 1994;269:31620–31625.

68 Lindh E, Bjork I: Binding of secretory component to dimers of immunoglobulin A in vitro. Eur J Biochem 1974;45:261–268.

69 Cunningham-Rundles C, Lamm ME: Reactive half cystine peptides of the secretory component of human exocrine immunoglobulin A. J Biol Chem 1975;250:1987–1991.

70 Fallgreen–Gebauer E, Gebauer W, Bastian A, Kratzin HD, Eiffert H, Zimmermann B, Karas M, Hilschman N: The covalent linkage of secretory component to IgA. Biol Chem Hoppe–Seyler 1993; 374:1023–1028.

71 Chintalacharuvu KR, Tavill AS, Loui LN, Vaerman JP, Lamm ME, Kaetzel CS: Disulphide bond formation between dimeric IgA and the polymeric immunoglobulin receptor during hepatic trans-cytosis. Hepatology 1994;19:162–173.

72 Burnet RC, Hanly WC, Zhai SK, Knight KL: The IgA heavy chain gene family in rabbit: Cloning and sequence analysis of 13 Cα genes. EMBO J 1989;8:4041–4047.

73 Liew TN, Russell SM, Appleyard G, Brand CM, Beale J: Cross-protection in mice infected with influenza virus is correlated with local IgA activity rather than serum antibody or cytotoxic T cell reactivity. Eur J Immunol 1984;14:350.

74 Tarkowski A, Lue C, Moldoveneau Z, Kiyono H, McGhee JR, Mestecky J: Immunization of humans with polysaccharide vaccines induces systemic predominantly IgA2-subclass antibody responses. J Immunol 1990;144:3770–3778.

75 Hajishengallis G, Nikolova E, Russell MW: Inhibition of *Streptococcus mutans* adherence to saliva coated hydroxy-apatite by human secretory antibodies to cell surface protein antigen I/II: Reversal by IgA protease cleavage. Infect Immunol 1992;60:5057–5064.

76 Grzych JM, Grezel D, Xu C, Neyrinck JL, Capron M, Ouma JH, Butterworth AE, Capron A: IgA antibodies to a protective antigen in human *Schistosoma mansoni*. J Immunol 1993;150:527–535.

77 Carayannopoulos L, Hexham JM, Capra JD: Localization of the binding site for the monocyte IgA-Fc receptor (CD89) to the domain boundary between Cα2 and Cα3 in human IgA1. J Exp Med 1996;183:1579–1586.

78 Woof JM, Partridge LJ, Jefferis R, Burton DR: Localisation of the monocyte binding region on human IgG. Mol Immunol 1986;23:319–330.

79 Nissim A, Eshhar Z: The human mast cell receptor binds to the third constant domain of immuno-globulin E. Mol Immunol 1992;29:1065–1072.

80 Deisenhofer J: Crystallographic refinement and atomic models of a human Fc fragment and its complex with fragment B of protein A from *Staphylococcus aureus* at 2.9 and 2.8 ångstroms resolution. Biochemistry 1981;20:2361–2370.

81 Kim JK, Tsen MF, Ghetie V, Ward ES: Localization of the site of the murine IgG1 molecule that is involved in binding to the murine intestinal Fc receptor. Eur J Immunol 1994;24:2429–2434.

82 Kim JK, Tsen MF, Chetie V, Ward ES: Identifying amino acid residues that influence plasma clearance of murine IgG1 fragments by site directed mutagenesis. Eur J Immunol 1994;24:542–548.

83 Schneiderman RD, Lint TF, Knight KL: Activation of the alternative pathway of complement by twelve different mouse-rabbit chimeric transfectoma IgA isotypes. J Immunol 1990;145:233–237.

84 Rits M, Hiemstra PS, Bazin H, van Es LA, Vaerman JP, Daha MR: Activation of rat complement by soluble and insoluble rat IgA immune complexes. Eur J Immunol 1988;18:1873–1880.

85 Mazanec MB, Nedrud JG, Kaetzel CS, Lamm ME: A three-tiered view of the role of IgA in mucosal defense. Immunol Today 1993;14:430–435.

86 Kaetzel CS, Robinson JK, Chintalacharuvu KR, Vaerman JP, Lamm ME: The polymeric immuno-globulin receptor (secretory component) mediates transport of immune complexes across epithelial cells: A local defense function for IgA. Proc Natl Acad Sci USA 1991;88:8796–8800.

87 Kaetzel CS, Robinson JK, Lamm ME: Epithelial transcytosis of monomeric IgA and IgG cross linked through antigen to polymeric IgA. J Immunol 1994;152:72–78.

88 Vaerman JP, Heremans JF, van Kerckhoven G: The identification of IgA in several mammalian species. J Immunol 1969;103:1421–1423.

89 Bell RB, Stephens CJ, Turner KJ: Marsupial immunoglobulins: An immunoglobulin resembling eutherian IgA in serum and secretions. J Immunol 1974;113:371–378.

90   Lebacq-Verheyden AM, Vaerman JP, Heremans JF: A possible homologue of mammalian IgA in chicken serum and secretions. Immunology 1972;22:165–175.

91   Mansikka A: Chicken IgA H chains: Implications concerning the evolution of H chain genes. J Immunol 1992;149:855–861.

92   Portis JL, Coe JE: IgM the secretory immunoglobulin of reptiles and amphibians. Nature 1975; 258:547–548.

93   Lobb CJ, Clem LW: Phylogeny of immunoglobulin structure and function XI: Secretory immunoglobulins in the cutaneous mucus of sheepshead, *Achosargus probatocephalus*. Dev Comp Immunol 1981;5:587–596.

94   Burnet RC, Hanly WC, Zhai SK, Knight KL: The IgA heavy chain gene family in rabbit: Cloning and sequence analysis of 13 Ca genes. EMBO J 1989;8:4041–4047.

95   Schneiderman RD, Hanly WC, Knight KL: Expression of 12 rabbit IgA Ca genes as chimeric rabbit-mouse IgA antibodies. Proc Natl Acad Sci USA 1989;86:7561–7565.

96   Spieker-Polet H, Yam P-C, Knight KL: Differential expression of 13 IgA heavy chain genes in rabbit lymphoid tissues. J Immunol 1993;150:5457–5465.

97   Plaut AG: The IgA1 proteases of pathogenic bacteria. Annu Rev Microbiol 1983;37:603–622.

98   Max EE, Korsmeyer SJ: Immunoglobulin J chain gene. Structure and expression in B lymphoid cells. J Exp Med 1985;161:832–849.

99   Matsuuchi L, Cann GM, Koshland ME: Immunoglobulin J chain gene from the mouse. Proc Natl Acad Sci USA 1986;83:456–460.

100   Kulseth MA, Rogne S: Cloning and characterization of the bovine immunoglobulin J chain cDNA and its promoter region. DNA Cell Biol 1994;13:37–42.

101   Hughes GJ, Frutiger S, Paquet N, Jaton JC: The amino acid sequence of rabbit J chain in secretory immunoglobulin A. Biochem J 1990;271:641–647.

102   Mikoryak CA, Margolies MN, Steiner LA: J chain in *Rana catesbeiana* high molecular weight Ig. J Immunol 1988;140:4279–4285.

103   Takahashi T, Iwase T, Takenuchi N, Saito M, Kobayashi K, Moldoveanu Z, Mestecky J, Moro I: The joining (J) chain is present in invertebrates that do not express immunoglobulins. Proc Natl Acad Sci USA 1996;93:1886–1891.

104   Morton HC, Atkin JD, Owens RJ, Woof JM: Purification and characterisation of chimeric human IgA1 and IgA2 expressed in COS and chinese hamster ovary cells. J Immunol 1993;151:4743–4752.

105   Rindisbacher L, Cottet S, Witteek R, Kraehenbuhl J-P: Production of human secretory component with dimeric IgA binding capacity using viral expression systems. J Biol Chem 1995;270:14220–14228.

106   Ma JKC, Hiatt A, Hein M, Vine ND, Wang F, Stabila P, van Dolleward C, Mostov KE, Lehner T: Generation and assembly of secretory antibodies in plants. Science 1995;268:716–719.

107   Lo D, Pursel V, Linton PJ, Sandgren E, Behringer R, Rexroad C, Palmiter RD, Brinster RL: Expression of mouse IgA by transgenic mice pigs and sheep. Eur J Immunol 1991;21:1001–1006.

108   Mestecky J, McGhee JR: Prospects for human mucosal vaccines. Adv Exp Med Biol 1992;327:13–23.

109   Czerkinsky C, Quiding M, Eriksson K, Nordstrom I, Lakew M, Weneras CAK, Bjorck S, Svennerholm AM, Butcher E: Induction of specific immunity at mucosal surfaces: Prospects for vaccine development. Adv Exp Med Biol 1995;371B:1409–1416.

110   Chen SY, Bagley J, Marasco WA: Human Gene Therapy 1994;5:595–601.

111   Ferkol T, Kaetzel CS, Davis PB: Recombinant human respiratory sycytial virus (RSV) monoclonal antibody Fab is effective therapeutically when introduced directly into the lungs of RSV-infected mice. J Clin Inv 1993;92:2394–2400.

112   Burton DR: Monoclonal antibodies from combinatorial libraries. Accounts Chem Res 1993;26: 405–411.

113   Winter G, Griffiths AD, Hawkins RE, Hoogenboom HR: Making antibodies by phage display technology. Annu Rev Immunol 1994;12:433–455.

J. Mark Hexham, Molecular Immunology Center, Department of Microbiology,
UT Southwestern Medical Center at Dallas, 6000 Harry Hines Boulevard,
Dallas, TX 75235–9140 (USA)

Capra JD (ed): Antibody Engineering.
Chem Immunol. Basel, Karger, 1997, vol 65, pp 88–110

....

IgG Effector Mechanisms

Michael R. Clark

Immunology Division, Department of Pathology, Cambridge University,
Cambridge, UK

Antibodies, otherwise known as immunoglobulins, are glycoproteins which perform a major role in helping to defend the host against infection. They exist as both membrane-bound receptors of B lymphocytes and also as secreted proteins, making up about 10–20% of plasma protein concentration in most mammalian species [1, 2]. Immunoglobulins are found in most vertebrates including fishes, amphibians, reptiles, birds and of course mammals. Their general structure has been conserved through evolution, being made up of a combination of 'heavy' and 'light' chains containing multiple homologous protein domains having a characteristic fold of a series of anti-parallel β-strands rolled up into a globular β-barrel structure stabilised by a conserved disulphide bond between two of the strands [3]. Most immunoglobulins have a basic symmetrical structural subunit of two heavy chains and two light chains giving rise to two identical antigen-binding sites [1–4].

The N-terminal domain is called the variable or V region domain and it is somatic changes in and selection of this which give rise to differences in antibody-binding affinity and specificity for antigen. Immunoglobulins in most species also exist in several different classes and sometimes subclasses, characterised by different numbers and types of constant or C region domains associated with virtually any of the possible V regions [1–3]. The differences in the C regions or classes of the antibodies allow for antibodies with different effector functions [1–4].

Comparative studies between species show that although immunoglobulins exist in nearly all vertebrates there are differences in the types found and in their functional properties. In mammalian species, the immunoglobulin classes consist of IgM, IgD, IgG, IgA and IgE with heavy chains designated respectively as μ, δ, γ, α and ε [1–4]. Detailed studies within some species have revealed that sometimes these classes are further subdivided into subclasses, although the precise numbers of subclasses and the gene organisation vary between species, indicating that these have arisen as recent evolutionary events [5]. Thus in rabbits IgA exists as thirteen subclasses [6], whilst in humans there are two IgA subclasses and in rats and mice only one [1–3]. Many species also have several subclasses of IgG including humans, rats and mice which have four subclasses whilst rabbits only seem to have one IgG [1–5]. In addition there are two commonly occurring types of immunoglobulin light chain consisting of one V-region domain and one C region domain called κ and λ [1–3]. As a result of the processes which regulate somatic rearrangements and expression of these immunoglobulin genes, each single B cell shows isotypic and allotypic exclusion by producing antibodies with a single type and sequence of heavy chain associated with a single type and sequence of light chain [1, 2, 4]. The usage of κ and λ varies between species and in rats and mice approximately 90–95% of immunoglobulins have κ light chains and 5–10% λ, whereas in humans the figure is nearer 60% κ and 40% λ.

IgG Isotypes

The IgG class is one which is characteristic of mammalian species although both it and the mammalian IgE class seem to share some evolutionary characteristics with the amphibian and avian IgY class [7]. IgG is the major class of immunoglobulin found in mammalian sera and it has a heavy chain with a V-region domain followed by three C region domains $C\gamma1$, $C\gamma2$, $C\gamma3$ (fig. 1) [1–4]. Between the $C\gamma1$ and $C\gamma2$ domains there is a less conserved proline-rich stretch of amino acids which appears to confer flexibility on the molecule, called the hinge region (fig. 1, 2a). In most IgG isotypes, the two heavy chains are covalently joined by disulphide bonds between cysteine residues in the hinge region although the exact number and positions of disulphides vary with subclass and species [1–3]. Again depending upon the subclass, each light chain is also usually covalently bonded by a single disulphide to a heavy chain, sometimes to a residue in the γ-chain hinge region (at position 220 for human IgG1), and at other times to a cysteine residue in the N-terminal half of the $C\gamma1$ domain at position 131 or 132 [3] (fig. 1). Interestingly, in species of the camel family, investigators have found that in addition to conventional IgG molecules they also have subclasses of IgG which do not associate with light chain and seem to lack a $C\gamma1$ domain [15].

A number of available crystal structures (fig. 2) show that in the Fab, the heavy- and light- chain V-region domains pack closely together and the antigen binding specificity is formed by the close arrangement of three hypervariable

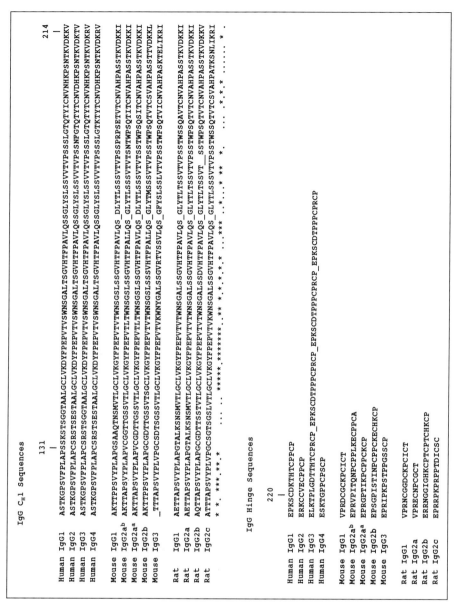

```
IgG CH1 Sequences

                   131                                                                                                                            214
                    |                                                                                                                              |
Human IgG1    ASTKGPSVFPLAPSSKSTSGGTAALGCLVKDYFPEPVTVSWNSGALTSGVHTFPAVLQSSGLYSLSSVVTVPSSSLGTQTYICNVNHKPSNTKVDKKV
Human IgG2    ASTKGPSVFPLAPCSRSTSESTAALGCLVKDYFPEPVTVSWNSGALTSGVHTFPAVLQSSGLYSLSSVVTVPSSNFGTQTYTCNVDHKPSNTKVDKTV
Human IgG3    ASTKGPSVFPLAPCSRSTSGGTAALGCLVKDYFPEPVTVSWNSGALTSGVHTFPAVLQSSGLYSLSSVVTVPSSSLGTQTYICNVNHKPSNTKVDKRV
Human IgG4    ASTKGPSVFPLAPCSRSTSESTAALGCLVKDYFPEPVTVSWNSGALTSGVHTFPAVLQSSGLYSLSSVVTVPSSSLGTKTYTCNVDHKPSNTKVDKRV

Mouse IgG1    AKTTPPSVYPLAPGSAAQTNSMVTLGCLVKGYFPEPVTVTWNSGSLSSGVHTFPAVLQS_DLYTLSSSVTVPSSPRPSETVTCNVAHPASSTKVDKKI
Mouse IgG2ab  AKTTAPSVYPLAPVCGGTTGSSVTLGCLVKGYFPEPVTLTWNSGSLSSGVHTFPALLQS_GLYTLSSSVTVTSNTWPSQTITCNVAHPASSTKVDKKI
Mouse IgG2aa  AKTTAPSVYPLAPVCGDTTGSSVTLGCLVKGYFPEPVTLTWNSGSLSSGVHTFPAVLQS_DLYTLSSSVTVTSSTWPSQSITCNVAHPASSTKVDKKI
Mouse IgG2b   AKTTPPSVYPLAPGCGDTTGSSVTSGCLVKGYFPEPVTVTWNSGSLSSSVHTFPALLQS_GLYTMSSSVTVPSSTWPSQTVTCSVAHPASSTVDKKL
Mouse IgG3    _TTAPSVYPLVPGCSDTSGSSVTLGCLVKGYFPEPVTVKNYGALSSGVRTVSSVLQS_GFYSLSSLVTVPSSTWPSQTVICNVAHPASKTELIKRI
              * *. ***.**.*      .. .: *****.*******..** *.** *.*.*.*    ..*.*  ...*** .*..:..* .: . .  :  .* . *

Rat IgG1      AETTAPSVYPLAPGTALKNSMVTLGCLVKGYFPEPVTVTWNSGALSSGVHTFPAVLQS_GLYTLTSSVTVPSSTWSSQAVTCNVAHPASSTKVDKKI
Rat IgG2a     AETTAPSVYPLAPGTALKSNSMVTLGCLVKGYFPEPVTVTWNSGALSSGVHTFPAVLQS_GLYTLTSSVTVPSSTWPSQTVTCNVAHPASSTKVDKKI
Rat IgG2b     AQTTAPSVYPLAPGCGDTTSSTVTLGCLVKGYFPEPVTVTWNSGALSSDVHTFPAVLQS_GLYTLTSSVT__SSTWPSQTVTCNVAHPASSTKVDKKV
Rat IgG2c     ATTTAPSVYPLVPGCSGTSGSLVTLGCLVKGYFPEPVTVKWNSGALSSGVHTFPAVLQS_GLYTLSSSVTVPSSTWSSQTVTCSVAHPATKSNLIKRI
              * *. ***.**.*       .. .:  *****.*******.***  *.**.*.*.*   ..***  .  * .  .: *.: * ..:::. *.

IgG Hinge Sequences

                 220
                  |
Human IgG1    EPKSCDKTHTCPPCP
Human IgG2    ERKCCVECPPCP
Human IgG3    ELKTPLGDTTHTCPRCP_EPKSCDTPPPCPRCP_EPKSCDTPPPCPRCP_EPKSCDTPPPCPRCP
Human IgG4    ESKTGPPCPSCP

Mouse IgG1    VPRDCGCKPCICT
Mouse IgG2ab  EPRVPITQNPCPLKECPPCA
Mouse IgG2aa  EPRGPTIKPCPPCKCP
Mouse IgG2b   EPSGPISTINPCPPCKECHKCP
Mouse IgG3    EPRIPKPSTPPGSSCP

Rat IgG1      VPRNCGGDCKPCICT
Rat IgG2a     VPRECNPCGCT
Rat IgG2b     ERRNGGIGHKCPCPTCHKCP
Rat IgG2c     EPRRPKRPPTDICSC
```

Fig. 1. Sequences for human, rat and mouse IgG subclasses derived from the Kabat database [3] are shown aligned for sequence homology. A '*' indicates complete homology between sequences with a '.' representing conservative changes, with spaces ' ' to allow better alignment. Residues mentioned in the text are indicated using the EU numbering system [3].

IgG C$_H$2 Sequences

```
              233              252          296            310      318      327
              | 235            |253         |297           |311     |320     |330
              |  | 238         ||254        ||             ||       || 322   || 331
              |  |  |          |||          ||             ||       |||      |||
Human  IGG1   APELLGGPSVFLFPPKPKDTLMISRTPEVTCVVVDVSHEDPEVKFNWYVDGVEVHNAKTKPREEQYNSTYRVVSVLTVLHQDWLNGKEYKCKVSNKALPAPIEKTISKAK
Human  IGG2   APP_VAGPSVFLFPPKPKDTLMISRTPEVTCVVVDVSHEDPEVQFNWYVDGVEVHNAKTKPREEQFNSTFRVVSVLTVVHQDWLNGKEYKCKVSNKGLPAPIEKTISKTK
Human  IGG3   APELLGGPSVFLFPPKPKDTLMISRTPEVTCVVVDVSHEDPEVQFKWYVDGVEVHNAKTKPREEQYNSTFRVVSVLTVLHQDWLNGKEYKCKVSNKALPAPIEKTISKTK
Human  IGG4   APEFLGGPSVFLFPPKPKDTLMISRTPEVTCVVVDVSQEDPEVQFNWYVDGVEVHNAKTKPREEQFNSTYRVVSVLTVLHQDWLNGKEYKCKVSNKGLPSSIEKTISKAK

Mouse  IgG1   VPEV___SSVFIFPPKPKDVLTITLTPKVTCVVVDISKDDPEVQFSWFVDDVEVHTAQTQPREEQFNSTFRSVSELPIMHQDWLNGKEFKCRVNSAAFPAPIEKTISKTK
Mouse  IgG2ᵇ  APDLLGGPSVFIFPPKIKDVLMISLSPMVTCVVVDVSEDDPDVQISWFVNNVEVHTAQTQTHREDYNSTLRVVSALPIQHQDWMSGKEFKCKVNNRALPSPIEKTISKPR
Mouse  IgG2aᵃ APNLLGGPSVFIFPPKIKDVLMISLSPIVTCVVVDVSEDDPDVQISWFVNNVEVHTAQTQTHREDYNSTLRVVSALPIQHQDWMSGKEFKCKVNNKDLPAPIERTISKPK
Mouse  IgG2b  APNLEGGPSVFIFPPNIKDVLMISLTPKVTCVVVDVSEDDPDVQISWFVNNVEVHTAQTQTHREDYNSTIRVVSTLPIQHQDWMSGKEFKCKVNNKDLPSPIERTISKIK
Mouse  IgG3   AGNILGGPSVFIFPPKPKDALMISLTPKVTCVVVDVSEDDPDVHVSWFVDNKEVHTAWTQPREAQYNSTFRVVSALPIQHQDWMRGKEFKCKVNNKALPAPIERTISKPK

Rat    IgG1   GSEV___SSVFIFPPKPKDVLTITLTPKVTCVVVDISQNDPEVRFSWFIDDVEVHTAQTHAPEKQSNSTLRSVSELPIVHRDWLNGKTFKCKVNSGAFPAPIEKSISKPE
Rat    IgG2a  GSEV___SSVFIFPPKPKDVLTITLTPKVTCVVVDISQDDPEVHFSWFVDDVEVHTAQTRPPEEQFNSTFRSVSELPILHQDWLNGRITFRCKVTSAAFPSPIEKTISKPE
Rat    IgG2b  VPELLGGPSVFIFPPKPKDILLISQNAKVTCVVVDVSEEEPDVQFSWFVNNVEVHTAQTQPREEQYNSTFRVVSALPIQHQDWMSGKEFKCKVNNKALPSPIEKTISKPR
Rat    IgG2c  DDNL_GRPSVFIFPPKPKDILMILTPKVTCVVVDVSEEEPDVQFSWFVDNVRVFTAQTQPHEEQLNGTFRVVSTLHIQHQDWMSGKEFKCKVNNKDLPSPIEKTISKPR

              .***.***.  ** ** * *  .  ********.***.*. .*....*  .  * .**  * *** . * .*  .** .. .*.**..***
```

IgG C$_H$3 Sequences

```
                                                          433
                                                          | 435
                                                          || 436
                                                          |||
Human  IGG1   GQPREPQVVTLPPSRDELTKNQVSLTCLVKGFYPSDIAVEWESNGQPENNYKTTPPVLDSDGSFFLYSKLTVDKSRWQQGNVFSCSVMHEALHNHYTQKSLSLSPGK
Human  IGG2   GQPREPQVVTLPPSREEMTKNQVSLTCLVKGFYPSDIAVEWESNGQPENNYKTTPPMLDSDGSFFLYSKLTVDKSRWQQGNVFSCSVMHEALHNHYTQKSLSLSPGK
Human  IGG3   GQPREPQVVTLPPSREEMTKNQVSLTCLVKGFYPSDIAVEWESSGQPENNYNTTPPMLDSDGSFFLYSKLTVDKSRWQQGNIFSCSVMHEALHNRFTQKSLSLSPGK
Human  IGG4   GQPREPQVVTLPPSQEEMTKNQVSLTCLVKGFYPSDIAVEWESNGQPENNYKTTPPVLDSDGSFFLYSRLTVDKSRWQEGNVFSCSVMHEALHNHYTQKSLSLSLGK

Mouse  IgG1   GRPKAPQVVTIPPPKEQMAKDKVSLTCMITDFFPEDITVEWQWNGQPAENYKNTQPIMNTNGSYFVYSKLNVQKSNWEAGNTFTCSVLHEGLHNHHTEKSLSHSPGK
Mouse  IgG2aᵇ GSVRAPQVVLPPPAAEEMTKKEEFSLTCMITGFLPAEIAVDWTSNGRTEQNYKNTATVLDSDGSYFMYSKLRVQKSTWERGSLFACSVVHEVLHNHLTTKTISRSLGK
Mouse  IgG2aᵃ GSVRAPQVVLPPPEEEMTKKQVTLTCMVTDFMPEDIVEWTNNGKTELNYKNTEPVLDSDGSYFMYSKLRVEKKNWVERNSYSCSVVHEGLHNHHTKSFSRTPGK
Mouse  IgG2b  GLVRAPQVVTLPPPAEQLSRKDVSLTCLVVGFNPGDISVEWTSNGHTEENYKDTAPVLDSDGSYFIYSKLNMKTSKWEKTDSFSCNVREGLKNYYLKKTISRSPGK
Mouse  IgG3   GRAQTPQVTIPPPREQMSKKKKVSLTCLVTNFFSEAISVEWERNGELEQDYKNTPPILDSDGTYFLYSKLTVDTDSWLQGEIFTCSVVHEALHNHHTQKNLSRSPGK

Rat    IgG1   GRTQVPHVTMSPTKEEMTQNEVSITCMVKGFYPPDIVEWQMNGQPQQENYKNTPPTMDTDGSYFLYSKLNVKKEKWQQGNTFTCSVLHEGLHNHHTEKSLSHSPGK
Rat    IgG2a  GTPRGPQVVTMAPPKEEMTQSQVSITCMVKGFYPPDIYTEWKMNGQPQENYKNTPPTMDTDGSYFLYSKLNVKKETWQQGNTFTCSVLHEGLHNHHTEKSLSHSPGK
Rat    IgG2b  GLVRKPQVVVMGPPTEQLTEQTVSLTCLTSGFLPNDIGVEWTSNGHIEKNYKNTEPVMDSDGSFFMYSKLNVERSRWDSRAPPVCSVVHEGLHNHHVEKSIRPPGK
Rat    IgG2c  GKARTPQVVTIPPPREQMSKNKVSLTCMVTSFYPASISVEWERNGELEQDYKNTLPVLDSDESYFLYSKLSVDTDSWMRGDIYTCSVVHEALHNHHTQKNLSRSPGK

              * .*.**.*   .**.  .*   *.*   .  .*.  .    * .*  *     . . ** .*  * **** **.* .  ** .*.*.
```

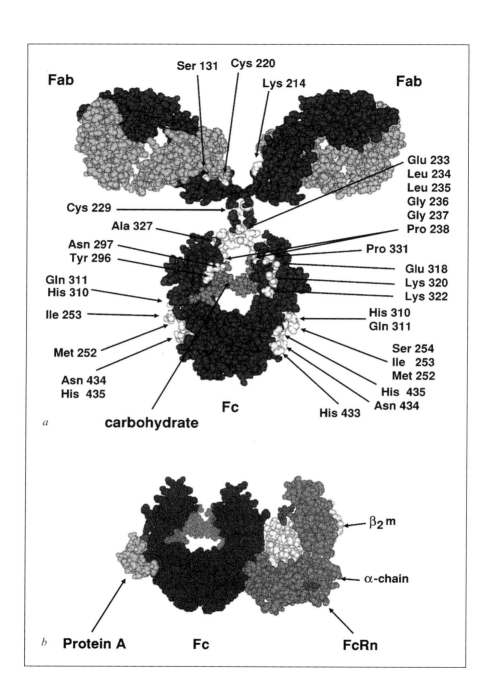

Fab

Ser 131   Cys 220
              Lys 214

Fab

Glu 233
Leu 234
Leu 235
Gly 236
Gly 237
Pro 238

Cys 229

Ala 327

Asn 297
Tyr 296

Pro 331

Gln 311
His 310

Glu 318
Lys 320
Lys 322

Ile 253

His 310
Gln 311

Met 252

Ser 254
Ile   253
Met 252

Asn 434
His 435

His 435
Asn 434

His 433

Fc

a   carbohydrate

β₂m

α-chain

b   Protein A   Fc   FcRn

loops from each domain [8]. The Cγ1 domain also has a close packing arrangement with light-chain Cκ or Cλ domain [8]. In the Fc fragment the two Cγ3 domains also show a tight packing arrangement, but in contrast, the two Cγ2 domains are held apart with the space between being occupied by conserved N-linked carbohydrate on each of the chains [9]. The two heavy chains are again brought together at the N-terminal end of the Cγ2 domains by the covalent disulphide bonding of the hinge region.

As can be seen from the human, rat and mouse sequences shown in figure 1, the IgG molecule is highly homologous between species. The four human IgG subclasses are very similar in sequence (greater than 90% over the C-region domains Cγ1, Cγ2 and Cγ3) probably as a result of recent gene duplication events within evolution [3, 4, 16, 17]. As will be described below, this high homology within human IgG subclasses has enabled the important residues affecting function to be studied [4, 16–19].

IgG Functions

Essentially all antibodies are able to act as adaptable linkers to the available effector systems by virtue of the ability to recognise and bind antigen through the V regions whilst interacting with conserved effector systems through the C-regions. Having two identical binding sites in a single IgG molecule allows the antibody to bind with higher avidity to antigens with repeating epitopes or to aggregates of antigen [1, 2, 4]. The flexibility of the IgG molecule, inherent within the elbow bend of the Fab (between V and C domains), and particularly within the hinge region allows for the two binding sites to cope with antigenic epitopes with a range of spacing and orientations (fig. 2) [1, 2, 4, 13, 20]. The binding of antigen can result in direct inactivation of infectious agents by blocking of functional sites with enzymic or receptor-binding activity. More importantly, IgG antibodies aggregated by antigen can interact with other

Fig. 2. Shown are raster space filling models of a human IgG1 antibody *(a)* and the Fc of an IgG antibody interacting with *S. aureus* protein A on the left and FcRn on the right. *(b)* The models were created from PDB structure files in the Brookaven Database [8] for a human Fab [9], a human IgG1 Fc [10], a human IgG1 Fc-protein A complex [11], and a human IgG1-rat FcRn complex [12]. Certain residues, in particular parts of the hinge, are missing from the crystal structures although attempts have been made to computer-model these features [13, 14]. For the models shown here, a peptide for the missing residues in the lower hinge region was created and inserted between the Fab and Fc structures. The final models were then aligned, superimposed and energy-minimilised for the protein backbones using the computer software packages Quanta and Charmm, from Biosym Technologies Inc. USA, running on a Silicon Graphics Iris Indigo workstation.

components of the immune system by either activating the complement cascade or by binding to Fcγ receptors on various cell types [1, 2, 4]. Both of these processes can assist in the opsonisation of antigen and in the triggering of inflammation and the enhancement of an immune response against an infectious agent [1, 2]. In addition to these antigen-dependent functions, IgG is also transported specifically to the neonate from the maternal plasma and also IgG antibodies show a reduced catabolic rate (or increased half-life) compared to other immunoglobulin classes and plasma proteins [1, 2, 4, 21]. Certain bacteria have exploited conserved sequences on the IgG molecule and have evolved proteins, e.g. *Staphylococcus aureus* protein A and streptococcal protein G, with high affinity for many IgG subclasses from many different species (fig. 2b) [1, 2, 4, 22]. In addition, low-affinity auto-antibodies termed 'rheumatoid factors', which are often IgM anti-IgG Fc antibodies, have been identified in several species. The sites recognised by these antibodies appear to be conserved and many of these rheumatoid factor antibodies also make use of conserved germline V-region sequences [23].

Yet another property reported for IgG subclasses in certain species (principally mouse IgG3 and perhaps the homologous rat IgG2c) is an ability to self-aggregate through sites in the Cγ3 region of the Fc [24]. Thus when these IgG antibodies bind to antigen their Fc regions allow large aggregates to form. Monoclonal antibodies of these subclasses also have a tendency to aggregate when purified, particularly when stored in the cold (cryoglobulins) or at lower ionic strength (euglobulins).

In species such as the rabbit, the single IgG class of antibody seems able to activate complement as well as to bind to Fcγ receptors, however in several other species, gene duplication events have led to multiple IgG subclasses (e.g. human, rat and mouse having 4 IgG subclasses each) and then through changes in sequence some of these subclasses have lost the ability to activate complement or to bind to some FcγR types [1–5, 16–18]. Although IgG antibodies and their associated effector systems are obviously conserved and homologous, between species the numbers of IgG subclasses vary and many of the duplications seem to have occurred after speciation [4, 5]. This suggests that there may not be direct functional equivalence between subclasses in each species.

Any consideration of the interactions of IgG antibodies with Fc receptors needs to take account of the different classes of receptor. In humans, there are three identified classes of Fc receptors for human IgG (FcγR). Much is known about the gene organisation and the structure and function of these receptors and they can be readily detected on the cell surface using specific monoclonal antibodies [25, 26]. Human FcγRI (CD64) can bind monomeric IgG with high affinity and is expressed constitutively on macrophages and monocytes and can be induced on neutrophils and eosinophils. Human FcγRII

(CD32) binds IgG only in complexed or polymeric forms and is widely expressed on a range of cell types including monocytes, macrophages, basophils, eosinophils, Langerhans cells, B cells and platelets. Human FcγRIII (CD16) is also a medium- to low-affinity receptor and is expressed on macrophages, some large granular lymphocytes (LGL), killer cells (K cells), some natural killer cells (NK cells), and neutrophils and can be induced on eosinophils and monocytes [25, 26].

However, the expression of these three receptor classes on cell types is very complex for several reasons [25, 26]. Firstly, each of these three classes of receptor is encoded by a cluster of closely related genes, FcγRIA, B and C, FcγRIIA, B and C and FcγRIIIA and B. Most of these genes encode transmembrane receptors which either have cytoplasmic domains capable of signal transduction or which associate with signalling co-receptor complexes [26]. For example, the FcγRIIIa form is a transmembrane receptor expressed in conjunction with γ-, ζ- and β-chains, and is found on K/NK cells, monocytes and macrophages. This receptor complex is very homologous to the T cell receptor complex with CD3. In contrast to the other Fcγ receptors, the human (but not the mouse) FcγRIIIB, found on neutrophils and eosinophils is not transmembrane but is instead a glycosyl-phosphatidyl-inositol (GPI)-anchored receptor [25, 26]. Additional complexity arises because some of these genes then give rise to multiple transcripts, e.g. FcγRIIb1, b2 and b3 and finally some of these genes exist within the population in different polymorphic forms, e.g. FcγRIIa-R131/FcγRIIa-H131 and FcγRIIIbNA1/FcγRIIIbNA2 [25].

Recombinant IgG Antibodies

The recognition that the specificity resides within the variable domains of the IgG molecule whilst the functions of the different subclasses reside within the constant domains facilitates the ready manipulation of the genes and hence the proteins as recombinant products. Thus V regions from mouse or rat monoclonal antibodies were co-expressed with human IgG C regions to create chimeric antibodies with human effector functions [28–32]. Such molecules were recognised to be of potential therapeutic value and facilitate studies of the structural basis of antibody functions [31–33].

Further improvements in the protein engineering of recombinant antibodies involved manipulating sequences within individual domains in order to investigate the structural basis for IgG effector functions. Two basic approaches have been used. Firstly researchers have compared the functions and sequences of naturally occurring IgG molecules of different subclasses and from different species [3, 16, 29–32, 34]. These comparisons are used to make predictions about which sequences are involved in determining the functional differences. The predictions can be tested either by mutating residues to an inappropriate type such as to alanine or alternatively sequences in one IgG subclass can be changed to be similar to another subclass either by site-directed mutagenesis or by making chimeric constructs between the two subclasses [35–37].

Information on the effector functions of different IgG subclasses was confirmed using a number of matched panels of recombinant antibodies [29–31]. A complete matched set of monoclonal antibodies including the human IgG isotypes IgG1 [allotype G1m(za)], IgG2, IgG3 [allotypes G3m(b) and G3m(g)] and IgG4, was constructed with specificity for the hapten NP and its derivative NIP (5-iodo-4-hydroxy-3-nitrophenylacetyl) [30]. The antibodies were tested for their abilities to mediate autologous complement-dependent lysis of human red blood cells labelled with NIP, and for antibody-dependent cell-mediated cytotoxicity (ADCC) of a NIP-conjugated human lymphoblastoid cell line by activated human mononuclear cells. The IgG1 antibody proved to be the most effective in both complement-dependent and cell-mediated cytotoxicity [30]. The effector cells in ADCC were inhibited by a CD16 (FcγRIII) monoclonal antibody and had the phenotype of K cells. The two IgG3 antibodies were not quite as good as the IgG1 antibody in either assay and the two allotypes of IgG3 showed differences in complement lysis. Surprisingly, this was despite the fact that IgG3 was shown to be fixing many more molecules of C1q than IgG1. Further studies demonstrated that although human IgG1 bound less C1 than human IgG3, there was a much more efficient deposition of C4b on the cell surface which accounted for the more effective cell lysis by IgG1 [38]. The other isotypes, IgG2 and IgG4, did not show any significant functional activity in these assays [30].

Two independent groups have investigated the effect of varying the antigen density as well as the epitope patchiness for complement lysis triggered by the NIP chimerics [39, 40]. The IgG1 antibody was most effective when the antigen concentration was higher, whereas the IgG3 antibody was relatively better at lower concentrations. The IgG2 antibody gave good lysis at very high concentrations of antigen but the IgG4 antibody did not give lysis under any conditions. It was also shown that IgG1 and IgG3 activated the classical pathway of complement, but not the alternative pathway. At high antigen concentration the IgG2 antibody could also activate the classical pathway, but the IgG4 antibodies could not. However, the IgG2 antibody, did activate the alternative pathway of complement.

In another series of studies using matched sets of human IgG1, 2, 3 and 4 antibodies to the conventional human cell surface antigen CD52, the IgG1 antibody was again found to be the most effective antibody in complement-mediated lysis [32]. The IgG3 antibody was the next most active (two-fold lower in titre) followed by the IgG2 antibody (10-fold lower titre) whilst the IgG4 antibody did not work at all in complement lysis. Comparing these results with the NIP chimerics revealed a large degree of similarity except that the IgG2 antibody to CD52 antigen was more active in complement lysis [30, 32]. As described above, this observation may be due to different antigen densities, or clustering of antigen, the CD52 antigen being equivalent to the NIP antigen at high concentrations [39, 40].

Residues Affecting Complement Binding

As mentioned above, human IgG4 is the only human IgG subclass which does not activate complement, and the subclasses IgG1 and 3 are the most effective [16, 30, 32, 39, 40]. For the mouse, it is the subclasses IgG2a and IgG2b which are active with IgG1 and possibly IgG3 being inactive whilst in the rat, all four subclasses are active, with IgG2b and IgG1 being the most effective [16, 21, 34, 35]. Using comparisons of the amino acid sequences of these IgG antibodies, the residues which might be responsible for the different functions were identified (fig. 1, 2) [16]. This led to a series of experiments in which site-directed mutations of such residues were introduced into a comple-ment-fixing mouse IgG2b antibody, with specificity for the hapten NIP. A sequence motif in the $C\gamma2$ region involving Glu318, Lys320 and Lys322 was identified as being crucial for the binding of C1q the first component of the complement cascade [35].

From figure 1 it can be seen that this motif is also present in some isotypes which do not bind and activate complement such as human IgG4 and so there must be other features which are crucial to function. It was proposed that the difference between IgG3 and IgG4 was therefore dependent upon the former having a long and very flexible hinge region whilst the later has a short and more rigid hinge [16, 41]. Thus the greater segmental flexibility of the IgG3 might allow greater access to the Fc region of the antibody for complement binding. To answer this question, the hinges of IgG3 and IgG4 were swapped in anti-DNS antibodies but it was found that although the upper hinge length was responsible for the big differences in segmental flexibility, this was not responsible for the differences seen in complement binding [42]. Further experi-ments with domain swap mutants involving IgG1, 3 and 4 with specificity for the hapten DNS and then domain swaps involving IgG1 and IgG4 with specificity for the lymphocyte antigen CD52 demonstrated that the genetic hinge region only has a marginal effect on complement activation and the crucial differences between IgG4 and IgG1 are in the COOH terminal half of the $C\gamma2$ domain [43–45]. Between human IgG1 and IgG4 there are only four residues which differ in this region, (296 Tyr/Phe, 327 Ala/Gly, 330 Ala/Ser, 331 Pro/Ser), in the anti-DNS antibody system and also repeated with a NIP-hapten-specific antibody it was shown that a substitution of serine at position 331 in IgG4 for proline (as in IgG1, 2 and 3) endows IgG4 with the ability to activate complement [45, 46].

Recently, starting with a human IgG1 antibody specific for the human MHC class-II, HLA-DR antigen, mutations were introduced to attempt to alter complement and FcγR binding [47]. Surprisingly, two results with this human IgG1 antibody show a big difference with earlier studies using mouse IgG2b antibodies specific for NIP [35]. Firstly, a change in the residue 320, previously reported from experiments with mouse IgG2b as being a crucial residue for C1q binding, from Lys to Ala had no effect on complement. In contrast, a change in the residue Leu 235 to Glu, which had previously been implicated in FcγRI binding but not in complement activation using the mouse IgG2b, abolished complement lysis by the human IgG1 [36]. It should be noted that in the three-dimensional folding of the β-strands residues which are distant in the linear chain are brought closer together (fig. 1, 2).

The question of the role of the hinge in IgG is still not settled completely. In the experiments with anti-DNS, the hinge length of the IgG3 was shortened by deleting the repeat exons encoding a repeating peptide pattern (fig. 1). Whilst it was reported that an antibody without a hinge was inactive in complement activation, the shortened hinge versions were all apparently similarly active [42]. However, in a system using antibodies specific for the hapten NIP, it was reported that shortening the hinge of IgG3 resulted in an improvement in the complement activation by this isotype [48, 49]. Also in this NIP system, the hinge could be completely replaced with a single disulphide bond whilst still retaining the ability to activate complement [49, 50]. Human IgG1 has a different heavy-light chain disulphide bridge, compared to IgG2, 3 and 4, whereby the light chain is disulphide bonded to the upper hinge region of the antibody at position 220 as opposed to a cysteine normally found at position 131 in the first half of the Cγ1 domain (fig. 1). In the three-dimensional structure, these residues are about 0.6 nm apart (fig. 2). A set of mutated chimeric antibodies specific for the GD2, melanoma surface antigen, in which the arrangement of the light-heavy chain bond in IgG1 was altered to be more like the other subclasses, was prepared and compared in complement lysis with human complement [51]. The mutated IgG1 antibodies lost their ability to activate complement, indicating that residues in the Fab and upper hinge region are critical [51].

Further evidence for an involvement of residues in the Fab comes from recent studies on mutated human IgG1 antibodies to the CD52 antigen [33]. In human IgG1 there are several naturally occurring allotypes of the heavy chain in the human population which vary in frequency in different racial groups [52, 53]. Alternative allotypic residues are recognised at positions 214 in the Cγ1 (proximal to the hinge) and at positions in the Cγ3. A gene encoding the wild type allotype G1m(1,17) was mutated to give either the alternative natural allotype G1m(3), or artificial allotypes G1m(1,3) or G1m(17) [18, 33]. Surprisingly, it was found that the ability of the antibodies to cause complement-mediated lysis was quantitatively dependent on the residue found at position 214 (fig. 1, 2) in the Fab [18, 33].

Residues Affecting High-Affinity FcγRI (CD64) Binding

The same strategies outlined above can be used to determine residues critical for Fcγ receptor binding and activation. The high-affinity binding of monomeric IgG to the FcγRI receptor allowed this interaction to be studied. Human IgG1 and IgG3 bind FcγRI with the highest affinity (K_d 10^{-8} to 10^{-9} M) followed by IgG4 which is about 10-fold weaker in its interaction whilst IgG2 does not readily bind to the receptor [16, 25]. Rat IgG2b and mouse IgG2a also bind human FcγRI readily whilst mouse IgG2b antibodies were found to be poor at binding [54, 55]. A sequence comparison of IgG Classes indicated that residues in the IgG lower hinge region (encoded by the 5′ end of the Cγ2 exon) might be crucial (fig. 1, 2). Changing the residue Glu235 to Leu using site-directed mutagenesis improved the affinity of mouse IgG2b for FcγRI by 100-fold [36]. It was shown using reciprocal domain swap mutants between TNP specific IgG1 and IgG2 antibodies that a region spanning 233–238 (Glu Leu Leu Gly Gly Pro in IgG1) was critical for binding, and the introduction of the whole of this sequence into IgG2 produced an antibody which bound with higher affinity than IgG1 [56]. This result would indicate that there are of course other critical residues, possibly in the C-terminal half of the Cγ2 region but also perhaps the heavy-light chain disulphide bonding in the upper hinge region at position 220 (in IgG1) rather than 131 (in IgG2) which are critical. Domain swap mutants between DNS specific IgG2 and IgG3 antibodies also support the critical role of these residues encoded in the Cγ2 regions [57]. Investigation of the residues responsible for the lower binding affinity of IgG4 compared to IgG1 and 3 showed that a change of Phe234 in IgG4 to Leu improved the affinity [57]. However, in addition it was found that in IgG3 the change of Pro331 for Ser as found in IgG4 decreased the affinity of IgG3 for FcγRI [57]. In addition to these domain swap experiments, the introduction of point mutations into NIP specific IgG3 antibodies in which individual residues were changed to alanine indicated the critical role of the residues Leu234, Leu235 and Gly237 [58, 59].

Residues Affecting Low-Affinity FcγRII (CD32) Binding

Studies of the direct binding of IgG to the low affinity FcγRII are more difficult. Most experimental systems employ some form of complexed or aggregated IgG such as the rosetting of IgG-sensitised red cells by FcγRII-bearing leucocytes or alternatively IgG can be aggregated into small multimeric complexes using antigen or F(ab)$_2$ fragments of anti-light-chain-specific antibodies mixed in a one-to-one ratio with the IgG [60, 61]. CD3-specific antibodies are able to trigger a mitogenic response from T cells if the antibody Fc aggregates upon binding to Fc receptors on accessory cells and this can also be used to assay binding function to low-affinity receptors [55, 56]. Generally, the observations that IgG1 and IgG3 bound well to FcγRII but IgG4 and IgG2 did not again suggested that the sequence differences in the hinge and lower hinge regions might be critical (fig. 2). Using rosette formation between red cells sensitised by point-mutated IgG3 antibodies with the cell lines Daudi and K562 has indicated that some of the crucial residues for binding are Leu234 and Leu237 [59, 60].

A complication in the interpretation of these results is the observation that two alternative alleles of human FcγRIIa exist and that this leads to a functional difference in the ability of the receptor to discriminate between different IgG isotypes. This was originally identified for mitogenic responses with murine CD3 antibodies but has been found to affect binding of rat and human isotypes [54, 55]. Thus FcγRIIa-R131 binds mouse IgG1 but not human IgG2 whilst FcγRIIa-H131 does not bind mouse IgG1, but does bind human IgG2. Both forms of the receptor bind human IgG1, but do not bind human IgG4. Only a single amino acid change in the receptor (R→H131) is responsible for this remarkable ability to discriminate between the different isotypes. Further studies have also indicated that the polymorphism affects binding of rat IgG2b [55]. Rat IgG2b behaves in a similar way to human IgG2 and opposite to mouse IgG1; however, unlike human IgG2, the rat IgG2b also binds to the high-affinity FcγRI receptor [55]. Another observation made using mixed rat IgG2b with mouse IgG1 hybrid Fc antibodies was that two identical rat IgG2b heavy chains are required for binding to FcγRIIa but one is sufficient for measurable FcγRI binding [55].

Recent results suggest that functionally the FcγRIIa polymorphism may have important consequences with regard to resistance to certain infections as mediated by IgG2, and the racial differences in the allele frequencies may in part explain the observed geographical and racial differences in disease incidences such as *Haemophilus influenzae* infections [61]. In another clinical situation, this polymorphism in the binding of human IgG2 by FcγRIIa has been implicated as a risk factor in the development of heparin-induced thrombocytopenia [62].

Residues Affecting ADCC through FcγRIII (CD16)

As mentioned above, the FcγRIII receptor exists in two different forms in humans, either as a transmembrane receptor (FcγRIIIa) found on K and NK cells as well as activated macrophages or as a GPI anchored molecule (FcγRIIIb) on cells such as neutrophils. Most studies have concentrated on functional assays of ADCC using effector cells expressing the transmembrane FcγRIIIa. The rat IgG2b and the human IgG1 antibodies to NIP and to CD52 were particularly potent at triggering ADCC with human peripheral blood mononuclear cells and activated lymphocytes as the effectors whereas human IgG4 and rat IgG2a were poor [30, 32, 34, 63]. A series of domain swap mutants between human IgG1 and human IgG4 were constructed to identify residues responsible for the observed functional differences [44]. Firstly, it was verified that all of the critical differences between these two isotypes lie in the Cγ2 domain and secondly that they were in the COOH terminal end of the Cγ2 domain, a similar result to complement activation as described above and involving four possible amino acid changes (296 Tyr: Phe, 327 Ala:Gly, 330 Ala:Ser, 331 Pro:Ser) with in particular the residue change Pro331 in IgG1 for Ser in IgG4 prominent (fig. 1, 2) [44]. However in a different set of experiments using point mutations of residues in the lower hinge region of IgG3, the residues 235 and 237 were identified as critical for ADCC [59]. The importance of this region for ADCC through Fc$_\gamma$RIIIa was confirmed with the anti-HLA DR IgG1 antibody, where changing the Gly at 237 to Ala or exchanging the whole region 233 to 236 for the sequence found in IgG2 reduced activity [47]. As mentioned above in this same system a change of Leu235 to Glu not only abolished binding for FcγRI but also complement activation but had no effect on ADCC through FcγRIIIa [47]. The heavy-light chain disulphide bonding of IgG1 at position 220 in the hinge rather than 131 as in other classes may also be important. It was reported that the mutation of these residues in an IgG1 antibody specific for the melanoma antigen GD2 abolished ADCC by peripheral blood mononuclear cells although it was not formally demonstrated that FcγRIIIa was the receptor involved [51].

The results obtained with the domain swap mutants of IgG1 and IgG4 antibodies were however further complicated. It was found that the results obtained in this system were dependent upon the donor of the lymphocyte effectors [44]. With some donors, IgG1 was effective whilst IgG4 was ineffective, and in this case the domain swap mutants implicated the residues in the COOH half of the Cγ2 as critical. For some donors it was surprisingly found that IgG1 and IgG4 were both effective in ADCC and all of the domain swap mutants were indistinguishable. With such donors of effectors it was found that the four isotypes IgG1, IgG2, IgG3 and IgG4 gave very similar levels of activity [44]. Several genetic polymorphisms of the FcγRIIIa gene have recently been described, but it remains to be demonstrated which if any of these might be responsible for the functional polymorphism seen with the CD52 antibodies [64, 65].

As mentioned above, rat IgG2b antibodies were found to mediate ADCC with human K cells, whereas rat IgG2a antibodies did not [63]. In a similar fashion to the results with FcγRIIa, studies with hybrid Fc antibodies between rat IgG2b and rat IgG2a also showed that two identical rat IgG2b heavy chains were required in order to see functional ADCC [55, 63]. One explanation for these observations is that one IgG antibody needs to bind or interact with two FcγRIIa or FcγRIIIa receptors to mediate functional binding. Although it could be argued that this is due to an abnormal structure generated in the hinge region when two different isotypes are paired, this does not seem to disturb the interaction with FcγRI [55].

Glycosylation and Effector Functions

IgG molecules have a highly conserved N-linked glycosylation site within the Cγ2 domain at Asp 297 (fig. 1, 2) which has been found to be critical for complement-mediated lysis as well as binding to and activation of all three FcγR classes of receptor [1–4]. Antibodies produced without carbohydrate, either through use of metabolic inhibitors, endoglycosidases by site-directed mutation of the attachment site, or produced in bacterial expression systems, all show greatly reduced biological functions [16–18, 66, 67]. An aglycosylated human IgG antibody with specificity for the mouse CD8 antigen was found not to deplete mouse CD8 lymphocytes in vivo, whereas the glycosylated human IgG subclasses were very effective at depleting cells [68]. This property has been exploited in the production of an aglycosylated form of the humanised IgG1 CD3 antibody which can block T cell functions without depleting the cells or triggering cytokine release thus eliminating some of the severe side effects of CD3 antibody therapy [69]. Whilst glycosylation of antibody is important for function, the precise structures attached to the IgG are complex and can vary from one cell line to another depending upon the glycosyl transferases present [67, 70]. There is some suggestion that the precise carbohydrate structure present on an antibody might have some influence over the biological activity of the antibody in complement activation and FcγR binding although further investigation is revealing an important role for the precise oligosaccharide sequences and their interactions with the protein sequences found in each isotype [67, 71, 72]. Two functions of IgG antibody which do not seem to be strongly dependent upon glycosylation are neonatal transport and secondly the catabolic rate of IgG [21, 73].

Neonatal Transport and the Role of FcRn

An interesting aspect of the IgG class in mammals which have been studied is the acquisition of maternal IgG by the neonate [1, 2]. This is thought to provide a level of protection against infection early in life before the infants own adaptive immune response has had a chance to develop a significant repertoire of its own. However there are significant differences in the way that species obtain this IgG. In some species, for example rats, mice, horses and pigs, the IgG appears to be transferred to the infant during the first few hours of life in the colostrum and the IgG is then absorbed from the gut. In other species, humans being a key example, the IgG is transferred across the placenta to the neonate in utero.

An Fc receptor called FcRn was identified in rats and mice which seemed to be responsible for the transport of IgG across the gut during the first 24 h after birth. The interesting feature of this receptor is that it was found to be associated with β_2-microglobulin and had homology with MHC class I molecules [74]. Studies of β_2-microglobulin knockout transgenic mice have confirmed that it is required for the acquisition of maternal IgG by neonates but that IgG levels in maternal colostrum are unaltered [75]. The rat FcRn was co-crystallized with IgG and a full structure of the complex determined. The receptor has a structure very similar to MHC class I but does not appear to bind peptides in the groove between the two α-helices (fig. 2b) [76]. Instead, one end of the two α-helices seems to form a contact with the interface between the Cγ2 and Cγ3 domains of the IgG Fc [12]. Interestingly, this is a region very similar to a binding site in the Fc identified from a crystal structure of *S. aureus* protein A and human IgG1 (fig. 2b) [11]. Protein A binding shows a remarkable degree of conservation across species and between many IgG subclasses and perhaps the reason is that this same part of the IgG molecule is conserved through a requirement for interaction with FcRn-like receptors.

The receptor exhibits only weak binding with IgG at neutral pH, but shows increased affinity at low pH as might be encountered in the gut or in intracellular vesicles [77, 78]. It is thought that this pH dependence is involved in the recycling of the receptor and the reversible binding of FcRn to IgG. Using a soluble form of the FcRn receptor and recombinant IgG antibodies in plasmon resonance binding assays, the pH dependence was shown to be dependent upon titrating histidine residues in the Fc of the IgG molecule (residues 310 in the Cγ2 and 433 in the Cγ3 domain) and also in the FcRn heavy chain molecule itself (residues

250 and 251) (fig. 1, 2) [78]. Mutations in the Cγ2 residues 253, 310, 311, and the Cγ3 residues 433 and 434 of a mouse IgG1 antibody have been shown to reduce the transmission from the intestinal lumen of neonatal mice [21, 79]. These studies also investigated the effect of mutating one heavy chain only in the Fc pair and demonstrated that two functional heavy chains were required for efficient transport, and this suggests that transport involves two FcRn molecules interacting with one IgG [21, 79].

For many years, there has been a search for the human Fc receptor(s) responsible for the trans-placental transport of maternal IgG. Some workers have argued in favour of the receptors FcγRI, FcγRII and or FcγRIII as being involved in this transport. All four human IgG subclasses appear to be transported so this observation is not readily reconciled with the affinities of the four subclasses for the classical Fcγ receptors. A human homologue with high homology to the rat FcRn has been identified and sequenced, the molecule has an MHC-class-I-like α-chain associated with β_2-microglobulin [80]. The transfected and expressed molecule shows a similar pH dependence in binding to IgG as compared to rat FcRn. Also a recently published study using confocal fluorescent microscopy has identified β_2-microglobulin and IgG as being co-localized to intracellular apical granules in human placental syncytiotropho-blasts [81]. The evidence thus now seems compelling that a homologous MHC-class-I-like molecule is involved in maternal-fetal transport of IgG, the transport being across the gut in some animals and across the placenta in others.

Although acquisition of maternal IgG by the infant is normally thought to be protective, there are diseases and pathologies which result from the transport of inappropriate specificities. For example, haemolytic disease of the new-born occurs when antipaternal allotype blood group IgG antibodies (e.g. against Rh-D antigens in humans) are acquired by the infant, resulting in the destruction of red blood cells by an Fc dependent process [1, 2]. Other diseases are associated with maternal antibodies specific for platelet alloantigens (neonatal alloimmune thrombocytopenia) or, for example, autoantibodies against neurotransmitter receptors [82, 83]. Interestingly, haemolytic disease of the new-born is also observed in several domestic species, such as horses and pigs, but in line with the absence of placental transfer in these species the pathology only occurs when the infants suckle the colostrum.

Is IgG Catabolism Related to Binding to FcRn?

It has for many years been realised that the biological half-life of IgG molecules is abnormally long (3–4 weeks) when compared to other plasma proteins, including other immunoglobulin classes such as IgM and IgA (3–7 days) [1, 2,

21, 84]. The other property which has been noted is that the half-life is also dependent upon the total concentration of IgG in the plasma. Thus if the concentration of IgG is raised as in conditions such as myeloma, the half-life is greatly reduced whilst if the IgG concentration is lowered as in agammaglobunaemia, the half-life of administered immunoglobin is extended. This suggested a saturable receptor-driven mechanism for the control of the catabolic rate of IgG [85]. Thus a relatively constant plasma concentration of IgG can be maintained over a fairly wide range of synthetic rates. It was proposed that the IgG molecule might be endocytosed and then bound to a receptor which protected it from lysosomal degradation and then recycled it back to the plasma. There is currently strong evidence which indicates that the FcγRc (catabolism) receptor is similar or identical to the neonatal Fc receptor FcRn [19, 21].

Firstly, the half-life of IgG is largely a function of the Fc and in particular requires both the Cγ2 and Cγ3 domains [86]. Unlike the effector functions mediated through the receptors FcγRI, II and III the half-life is not drastically dependent upon glycosylation of the Cγ2 region, or upon amino acids implicated in binding to these three classes of receptors or to complement [21, 73]. Secondly conserved residues present within IgGs from most species and overlapping with the identified protein-A-binding site were identified as being important in controlling catabolic rate (residues 253, 310, 311, 433 and 434) [21, 87]. This is the same site to which the FcRn receptor binds as identified within the crystal structure [12]. In particular, the same histidine residues which have been identified as being important for the pH dependence of FcRn binding to IgG are also important for controlling catabolic rate [77, 78, 87]. This pH dependence of binding of FcRn would also fit with the recycling receptor hypothesis mentioned above [85]. Finally, the in vivo half-life of mouse IgG1 appears to require two heavy chains in the Fc which both have wild-type sequences just as for transport via the neonatal FcRn [21, 86].

So is FcRn or a closely related receptor, FcRc, responsible for the decreased catabolism of IgG molecules? A number of observations reported in the literature seem to implicate FcRn although the authors of these papers do not always make this a conclusion. Thus studies have shown (1) that the rat FcRn is expressed in adult hepatocytes and (2) in probing human tissues by Northern blot with the human homologue, expression of RNA was identified in a range of tissues including liver, kidney and pancreas [87, 88]. (3) It had been observed that the IgG concentration of the plasma of transgenic β₂m-deficient mice is lower than normal litter mates although the studies did not look at catabolic rates in these animals [75]. This definitive experiment has recently been published by Ghetie et al. [89] and they found that in β2m deficient mice the half-life of IgG1 was very short, suggesting that the one receptor FcRn is responsible for both neonatal transport and plasma IgG homeostasis [89].

There are some interesting predictions which might also relate to the role of the FcRn receptor in controlling catabolic rates. Human IgG3 tends to have a shorter half-life than the other human IgG subclasses the most common allotypic form has one of the histidine changes (His435) in the Cγ3 such that it does not bind to protein A [1, 2, 4, 16, 17, 21]. This might also affect the binding to the catabolic receptor and increase its catabolic rate. However, there is another allotype of IgG3 found mainly within the Japanese population which does have this histidine (435) and which also binds to protein A [4, 52]. An interesting question is whether these two allotypes would bind with different affinities and have different catabolic rates in vivo?

Acknowledgments

I would like to gratefully acknowledge the support of the Welcome Trust (grant reference number 034817/Z/91), and the earlier support of the UK Medical Research Council which enabled my own studies on the effector functions of IgG subclasses.

References

1   Janeway CA Jr, Travers P: Immunobiology: The immune system in health and disease, ed 2. Oxford, Blackwell Scientific Publications, 1996.
2   Male D, Cooke A, Owen M, Trowsdale J, Champion B: Advanced Immunology, ed 3. London, Times Mirror International Publishers, 1996.
3   Kabat EA, Wu TT, Perry HM, Gottesman KS, Foeller C: Sequences of proteins of immunological interest. Bethesda, US Department of Health and Human Services, NIH, 1991.
4   Clark M: General introduction; in Birch JS, Lennox E (eds): Monoclonal Antibodies: Principles and Applications. New York, Wiley, 1995, pp 1–43.
5   Bruggemann M: Evolution of the rat immunoglobulin gamma heavy chain gene family. Gene 1988; 74:473–479.
6   Schneiderman RD, Hanly WC, Knight KL: Expression of 12 rabbit IgA C genes as chimeric rabbit-mouse IgA antibodies. Proc Natl Acad Sci USA 1989;86:7561–7565.
7   Warr GW, Magor KE, Higgins DA: IgY: Clues to the origins of modern antibodies. Immunol Today 1995;16:392–398.
8   Bernstein FC, Koetzle TF, Williams GJB, Meyer EF Jr, Brice MD, Odger JR, Kennard O, Shimanou-chi T, Tasumi M: The protein data bank: A computer-based archival file for macromolecular structures. J Mol Biol 1977;112:535–542.
9   Saul FA, Amzel LM, Poljak RJ: Preliminary refinement and structural analysis of the Fab fragment from the human immunoglobulin NEW at 2.0 Å. J Mol Biol 1978;253:585–591.
10 Marquart M, Deisenhofer J, Huber R, Palm W: Crystallographic refinement and atomic models of the intact immunoglobulin molecule Kol and its antigen-binding fragment at 3.0 Å and 1.9 Å resolution. J Mol Biol 1980;141:369–375.
11 Deisenhofer J: Crystallographic refinement and atomic models of a human Fc fragment b of protein a from *Staphylococcus aureus* at 2.9 and 2.8 Å resolution. Biochemistry 1981;20:2361–2367.
12 Burmeister WP, Huber AH, Bjorkman PJ: Crystal structure of the complex of rat neonatal Fc receptor with Fc. Nature 1994;372:379–383.

13   Pumphrey RSH: Computer models of the human immunoglobulins. I. Shape and segmental flexibility. Immunol Today 1986;7:174–178.

14   Pumphrey RSH: Computer models of the human immunoglobulins. II. Binding sites and molecular interactions. Immunol Today 1986;7:206–211.

15   Hamers Casterman C, Atarhouch T, Muyldermans S, Robinson G, Hamers C, Songa EB, Bendahman N, Hamers R: Naturally occurring antibodies devoid of light chains. Nature 1993;363:446–448.

16   Burton DR: Immunoglobulin G: Functional sites. Mol Immunol 1985;22:161–206.

17   Jefferis R, Lund J: Molecular characterisation of IgG antibody effector sites; in Clark M (ed): Protein Engineering of Antibody Molecules for Prophylactic and Therapeutic Applications in Man. Nottingham, Academic Titles, 1993, pp 115–126.

18   Greenwood J, Clark M: Effector functions of matched sets of recombinant human IgG subclass antibodies; in Clark M (ed): Protein Engineering of Antibody Molecules for Prophylactic and Therapeutic Applications in Man. Nottingham, Academic Titles, 1993, pp 85–100.

19   Clark M: Unconjugated antibodies as therapeutics; in Harris WF, Adair J (eds): Antibody Therapeutics. Boca Raton, CRC Press, in press.

20   Pumphrey RSH: Introduction; in Shakib F (ed): The Human IgG SubClasses. New York, Pergamon Press, 1990, pp 3–14.

21   Ward ES, Ghetie V: The effector functions of immunoglobulins: Implications for therapy. Ther Immunol 1995;2:77–94.

22   Sauer-Eriksson AE, Kleywegt GJ, Uhel M, Jones TA: Crystal-structure of the C2 fragment of streptococcal protein-G in complex with the Fc domain of human-IgG. Structure 1995;3:265–278.

23   Jefferis R: Rheumatoid factors, B cells and immunoglobulin genes. Br Med Bull 1995;51:312–331.

24   Greenspan NS, Cooper LJN: Complementarity, specificity and the nature of epitopes and paratopes in multivalent interactions. Immunol Today 1995;16:226–230.

25   van de Winkel JGJ, Capel PJA: Human IgG Fc receptor heterogeneity: Molecular aspects and Clinical implications. Immunol Today 1993;14:215–221.

26   Ravetch JV: Fc Receptors: Rubor redux. Cell 1994;78:553–560.

27   Boulianne GL, Hozumi N, Schulman MJ: Production of functional chimaeric mouse/human antibody. Nature 1984;312:643–646.

28   Neuberger MS, Williams GT, Mitchell EB, Jouhal SS, Flanagan JG, Rabbitts TH: A hapten-specific chimeric immunoglobulin E antibody which exhibits human physiological effector function. Nature 1985;314:268–271.

29   Boulianne GL, Isenman DE, Hozumi N, Shulman MJ: Biological properties of chimeric antibodies: Interaction with complement. Mol Biol Med 1987;4:37–49.

30   Bruggemann M, Williams GT, Bindon CI, Clark MR, Walker MR, Jefferis R, Waldmann H, Neuberger MS: Comparison of the effector functions of human immunoglobulins using a matched set of chimeric antibodies. J Exp Med 1987;166:1351–1361.

31   Steplewski Z, Sun LK, Shearman CW, Ghrayeb J, Daddona P, Koprowski H: Biological activity of human-mouse IgG1, IgG2, IgG3, and IgG4 chimeric monoclonal antibodies with antitumor specificity. Proc Natl Acad Sci USA 1988;85:4852–4856.

32   Riechmann L, Clark MR, Waldmann H, Winter G: Reshaping human antibodies for therapy. Nature 1988;332:323–327.

33   Gorman SD, Clark MR: Humanisation of monoclonal antibodies for therapy. Semin Immunol 1990;2:457–466.

34   Bruggemann M, Teale C, Clark M, Bindon C, Waldmann H: A matched set of rat/mouse chimeric antibodies. Identification and biological properties of rat H chain constant regions μ, γ1, γ2a, γ2b, γ2c, ε, and α. J Immunol 1989;142:3145–3150.

35   Duncan AR, Winter G: The binding site for C1q on antibodies. Nature 1988;332:738–740.

36   Duncan AR, Woof JM, Partridge LJ, Burton DR, Winter G: Localization of the binding site for the human high-affinity Fc receptor on IgG. Nature 1988;332:563–564.

37   Shin S, Wright A, Bonagura V, Morrison SL: Genetically-engineered antibodies: Tools for the study of diverse properties of the antibody molecule. Immunol Rev 1992;130:8–107.

38   Bindon CI, Hale G, Bruggemann M, Waldmann H: Human monoclonal IgG antibodies differ in complement activating function at the level of C4 as well as C1q. J Exp Med 1988;166:351–356.

39   Valim YML, Lachmann PJP: The effect of antibody isotype and antigenic epitope density on the complement-fixing activity of immune complexes. Clin Exp Immunol 1991;84:1–8.

40   Michaelsen TE, Garred P, Aase A: Human IgG subclass pattern of inducing complement-mediated cytolysis depends on antigen concentration and to a lesser extent on epitope patchiness, antibody affinity and complement concentration. Eur J Immunol 1991;21:11–16.

41   Schneider WP, Wensel TG, Stryer L, Oi VT: Genetically engineered immunoglobulins reveal structural features controlling segmental flexibility. Proc Natl Acad Sci USA 1988;85:2509–2513.

42   Tan LK, Shopes RJ, Oi VT, Morrison SL: Influence of the hinge region on complement activation, C1q binding, and segmental flexibility in chimeric human immunoglobulins. Proc Natl Acad Sci USA 1990;87:162–166.

43   Tao M, Canfield SM, Morrison SL: The differential ability of human IgG1 and IgG4 to activate complement is determined by the COOH-terminal sequence of the C_H2 domain. J Exp Med 1991; 173:1025–1028.

44   Greenwood J, Clark M, Waldmann H: Structural motifs involved in human IgG antibody effector functions. Eur J Immunol 1993;23:1098–1104.

45   Tao MH, Smith RI, Morrison SL: Structural features of human immunoglobulin G that determine isotype specific differences in complement activation. J Exp Med 1993;178:661–667.

46   Brekke OH, Michaelsen TE, Aase A, Sin RH, Slie I: Human IgG isotype-specific amino acid residues affecting complement-mediated cell lysis and phagocytosis. Eur J Immunol 1994;24:2542–2547.

47   Morgan A, Jones ND, Nesbitt AM, Chaplin L, Bodmer MW, Emtage JS: The N-terminal end of the C_H2 domain of chimeric human IgG1 anti-HLA DR is necessary for C1q, FcγRI and FcγRIII binding. Immunology 1995;86:319–324.

48   Michaelsen TE, Aase A, Westby C, Slie I: Enhancement of complement activation and cytolysis of human IgG3 by deletion of hinge exons. Scan J Immunol 1990;32:517–528.

49   Brekke OH, Michaelsen TE, Slie I: The structural requirements for complement activation by IgG: Does it hinge on the hinge? Immunol Today 1995;16:85–90.

50   Brekke OL, Michaelsen TE, Sin R, Sanlie I: Activation of complement by an IgG molecule without a genetic hinge. Nature 1993;363:628–630.

51   Dorai H, Wesolowski JS, Gillies SD: Role of inter-heavy and light chain disulphide bonds in the effector functions of human immunoglobulin IgG1. Mol Immunol 1992;29:1487–1491.

52   WHO: Review of the notation for the allotypic and related markers of human immunoglobulins. J Immunogenet 1976;3:357–366.

53   van Loghem E: Allotypic markers; in Shakib F (ed): Basic and Clinical Aspects of IgG Subclasses. Monogr Allergy. Basel, Karger, 1986, vol 19, pp 40–51.

54   Tax WJM, Hermes FFM, Willems RW, Capel PJA, Koene RAP: Fc receptors for mouse IgG1 on human monocytes: Polymorphism and role in antibody-induced T cell proliferation. J Immunol 1984;133:1185–1189.

55   Haagen I-A, Geerars AJG, Clark MR, van de Winkel JGJ: Interaction of human monocyte Fcγ receptors with rat IgG2b: A new indicator for the FcγRIIA (R-H131) polymorphism. J Immunol 1995;154:1852–1860.

56   Chappel MS, Isenman DE, Everett M, Xu Y, Dorrington KJ, Klein MH: Identification of the Fc gamma receptor class I binding site in human IgG through the use of recombinat IgG1/IgG2 hybrid and point-mutated antibodies. Proc Natl Acad Sci USA 1991;88:9036–9040.

57   Canfield SM, Morrison SL: The binding affinity of human IgG for its high affinity Fc receptor is determined by multiple amino acids in the C_H2 domain and is modulated by the hinge region. J Exp Med 1991;173:1483–1491.

58   Jefferis R, Lund J, Pound J: Molecular definition of interaction sites on human IgG for Fc receptors (huFcγR). Mol Immunol 1990;27:1237–1240.

59   Sarmay G, Lund J, Rozsnyai Z, Gergely J, Jefferis R: Mapping and comparison of the interaction sites on the Fc region of the IgG responsible for the triggering antibody dependent cellular cytotoxicity (ADCC) through different types of human Fcγ receptor. Mol Immunol 1992;29:633–639.

60   Huizinga TWJ, Kerst M, Nuyens JH, Vlug A, von dem Borne AEG, Roos D, Tetteroo PAT: Binding characteristics of dimeric IgG subclass complexes to human-neutrophils. J Immunol 1989;142:2359.

---

61 Bredius RGM, de Vries CEE, Troelestra A, van Alphen L, Weening RS, van de Winkel JGJ, Out TA: Phagocytosis of *Staphylococcus aureus* and *Haemophilus influenzae* type B opsonised by polyclonal human IgG1 and IgG2 antibodies: Functional hFcγRIIa polymorphism to IgG2. J Immunol 1993;151:1463.

62 Burgess JK, Lindeman R, Chesterman CN, Chong BH: Single amino acid mutation of Fcγ receptor is associated with the development of heparin-induced thrombocytopenia. Br J Haematol 1995;91: 761–767.

63 Clark M, Waldmann H: T-cell killing of target cells induced by hybrid antibodies: Comparison of two bispecific monoclonal antibodies. JNCI 1987;79:1393–1401.

64 de Haas M, Koene HR, Kleijer M, de Vries E, Simsek S, van Tol MJD, Roos D, von dem Borne AEG: A triallelic FcγRIIIA polymorphism influences the binding of human-IgG by NK-cell FcγRIIIA. J Immunol 1996;156;2948–2955.

65 Vance BA, Huizinga TWJ, Wardwell K, Guyre PM: Binding of monomeric Human IgG defines an expression polymorphism of FcγRIII on large granular lymphocyte/natural killer cells. J Immunol 1993;151:6429–6439.

66 Nose M, Takano R, Nakamura S, Arata Y, Kyoguku M: Recombinant Fc of human IgG1 prepared in *Escherichia coli* system escapes recognition by macrophages. Int Immunol 1990;2:1109–1112.

67 Lund J, Takahashi N, Pound JD, Goodall M, Nakagawa H, Jefferis R: Oligosaccharide-protein interactions in IgG can modulate recognition by Fcγ receptors. FASEB 1994;9:115–119.

68 Isaacs JD, Clark MR, Greenwood J, Waldmann H: Therapy with monoclonal antibodies – an in vivo model for the assessment of therapeutic potential. J Immunol 1992;148:3062–3071.

69 Bolt S, Routledge E, Lloyd I, Chatenoud L, Pope H, Gorman SD, Clark M, Waldmann H: The generation of a humanised, non-mitogenic CD3 monoclonal antibody which retains in vitro immunosuppressive properties. Eur J Immunol 1993;23:403–411.

70 Lifely MR, Hale C, Boyce S, Keen M, Phillips J: Glycosylation and biological activity of CAMPATH-1H expressed in different cell lines and grown under different culture conditions, in press.

71 Wright A, Morrison SL: Effect of altered C_H2-associated carbohydrate structure on the functional properties and in vivo fate of chimeric mouse-human immunoglobulin G1. J Exp Med 1994;180: 1087–1096.

72 Jefferis R, Lund J, Goodall M: Recognition sites on human IgG for Fc gamma receptors: The role of glycosylation. Immunol Lett 1995;44:111–117.

73 Wawrzynczak EJ, Denham S, Parnell GD, Cumber AJ, Jones P, Winter G: Recombinant mouse monoclonal antibodies with single amino acid substitutions affecting C1q and high affinity Fc receptor binding have identical serum half-lives in the BALB/c mouse. Mol Immunol 1992;29:221.

74 Simister NE, Mostov KE: An Fc receptor structurally related to MHC class-I antigens. Nature 1989;337:184–187.

75 Israel EJ, Patel VK, Taylor SF, Marshak-Rothstein A, Simister NE: Requirement for a beta-2-microglobulin associated Fc receptor for acquisition of maternal IgG by fetal and neonatal mice. J Immunol 1995;154:6246–6251.

76 Burmeister WH, Gastinel LN, Simister NE, Blum ML, Bjorkman PJ: Crystal structure at 2.2 Å resolution of the MHC-related neonatal Fc receptor. Nature 1994;372:336–343.

77 Raghaven M, Chen MY, Gastinel LN, Bjorkman PJ: Investigation of the interaction between class I MHC-related Fc receptor and its immunoglobulin G ligand. Immunity 1994;1:303–315.

78 Raghavan M, Bonagura VR, Morrison SL, Bjorkman PJ: Analysis of the pH dependence of the neonatal Fc receptor/immunoglobulin G interaction using antibody and receptor variants. Biochemistry 1995;34:14649–14657.

79 Kim JK, Tsen MF, Ghetie V, Ward ES: Localization of the site of the murine IgG1 molecule that is involved in binding to the murine intestinal Fc receptor. Eur J Immunol 1994;24:2429–2434.

80 Story CM, Mikulska JE, Simister NE: A major histocompatibility complex class I-like Fc receptor cloned from human placenta: Possible role in transfer of immunoglobulin G from mother to fetus. J Exp Med 1994;180:2377–2381.

81 Kristoffersen EK, Matre M: Co-localization of beta 2-microglobulin and IgG in human placental syncytiotrophoblasts. Eur J Immunol 1996;26:505–507.

82 Newman PJ: Platelet GPIIb-IIIa: Molecular variations and alloantigens. Thromb Haemost 1991; 66:111–118.
83 Vincent A, Riemersma S, Hawke S, Newsomdavis J, Brueton L: Arthrogryposis associated with antibodies inhibiting fetal acetylcholine-receptor function. Ann Neurol 1994;36:325.
84 Waldman TA, Strober W: Metabolism of immunoglobulins: Progr Allergy, Basel, Karger, 1969, vol 13, pp 1–110.
85 Brambell FWR, Hemmings WA, Morris IG: A theoretical model of gamma-globulin catabolism. Nature 1964;203:1352–1355.
86 Kim JK, Tsen MF, Ghetie V, Ward ES: Catabolism of the murine IgG1 molecule – evidence that both C_H2-C_H3 domain interfaces are required for persistence of IgG1 in the circulation of mice. Scand J Immunol 1994;40:457–465.
87 Kim JK, Tsen MF, Ghetie V, Ward ES: Identifying amino-acid residues that influence plasma-clearance of murine IgG1 fragments by site-directed mutagenesis. Eur J Immunol 1994;24:542–548.
88 Blumberg RS, Koss T, Story CM, Barisani D, Polischuk J, Lipin A, Pablo L, Green R, Simister NE: A major histocompatibility complex I-related Fc receptor for IgG on rat hepatocytes. J Clin Invest 1995;95:2397–2402.
89 Ghetie V, Hubbard JG, Kim JK, Tsen MF, Lee Y, Ward ES: Abnormally short serum half-lives of IgG in beta-2-microglobulin-deficient mice. Eur J Immunol 1996;26:690–696.

Michael R. Clark, Immunology Division, Department of Pathology, Cambridge University, Tennis Court Road, Cambridge CB2 1QP (UK)

Capra JD (ed): Antibody Engineering.
Chem Immunol. Basel, Karger, 1997, vol 65, pp 111–128

....

Glycosylation of Antibody Molecules: Structural and Functional Significance

Roy Jefferis, John Lund

Department of Immunology, The Medical School, University of Birmingham,
Birmingham, UK

The IgG antibody molecule is a structural paradigm for members of the immunoglobulin superfamily. A majority of these molecules are glycoproteins and collectively they account for ∼ 70% of molecules currently undergoing development for possible in vivo therapeutic application. Whilst the oligosaccharide moiety of the IgG molecule accounts for only 2–3% of its mass, it has been shown to be essential for optimal activation of effector mechanisms leading to the clearance and destruction of pathogens. This suggests that glycosylation fidelity is an essential requirement of the IgG molecule and that it may be so for other recombinant glycoproteins produced by in vivo or in vitro techniques. Numerous studies have shown that whilst the defining biological activity of a glycoprotein molecule may not be dependent on its glycosylation, other essential characteristics are altered in aglycosylated forms, e.g. stability, pharmokinetics, antigenicity [1–3].

The IgG molecule is composed of three globular protein moieties, two Fabs and an Fc, that are linked through a flexible 'hinge' region that allows freedom for multiple spatial orientations of the globular moieties with respect to each other. A flexible upper hinge region provides mobility for the Fab regions and allows the paratope of each to bind its complementary epitope. A flexible lower hinge region similarly allows Fc mobility and accessibility within antigen/antibody complexes to engage one of a variety of effector activating ligands, e.g. Fcγ receptors, the C1 component of complement. A core hinge section is rich in proline and cystine residues, that form inter-heavy chain disulphide bridges, and has a rigid secondary structure.

Studies attempting to correlate physicochemical parameters with function were interpreted to suggest that the segmental flexibility of the hinge region

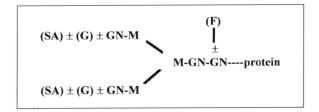

Fig. 1. The core carbohydrate moiety of the complex form of oligosaccharides is represented by the sugar residues in open type. The possible outer arm residues are bracketed.
All possible combinations are observed. SA = Sialic acid; G = galactose; GN = N-acetylglucosamine; M = mannose; F = fucose. N-linked attachment of oligosaccharide occurs on the
amide side chain of the Asn-x-Ser/Thr sequon (x ≠ Pro); the Ser/Thr residue forms hydrogen
bond(s) with the amide group in order to activate it for attachment to the primary N-
acetylglucosamine residue of the dolichol intermediate, by oligosaccharyltransferase.

was directly related to the ability of an IgG molecule to activate complement;
rather than indirectly by allowing access to interaction sites in the C_H2
and/or C_H3 domains for effector ligands [4]. The validity of these conclusions
has recently been re-evaluated by the application of protein engineering techniques in an attempt to introduce rational structural changes predicted to
affect biological activity. The results of these studies demonstrate the necessity
for Fc glycosylation and that protein/oligosaccharide interactions determine
the generation of a structure that is permissive of Fc-ligand recognition and
activation, while failing to confirm a primary role for the hinge region. Our
studies suggest that whilst the oligosaccharide moiety may not contribute
directly to ligand binding, except for mannan-binding protein, it does exert a
subtle influence on protein tertiary and quaternary structure that is essential to
'wild type' activity. Consequently, Fc-ligand recognition, and hence biological
activity, may be modulated by judicious replacement of amino acid residues
that contribute to non-covalent protein/oligosaccharide interactions.

Antibody Glycosylation

Human antibody molecules of the IgG class have N-linked oligosaccharide
attached at the amide side chain of Asn297 on the β-4 bend of the inner (Fx)
face of the C_H2 domain of the Fc region [5]. The oligosaccharide moiety is
of the complex biantennary type having a hexasaccharide 'core' structure
(GlcNAc2Man3GlcNAc) and variable outer arm 'non-core' sugar residues,
such as fucose, bisecting N-acetylglucosamine, galactose and sialic acid (fig. 1).

Thus, a total of 36 structurally unique oligosaccharide chains may be attached at each Asn297 residue. It is anticipated that glycosylation can be asymmetric so that an individual IgG molecule may have different oligosaccharide chains attached at each of the Asn297 residues within the Fc region such that whilst the heavy chain synthesised within a single antibody-secreting cell may be homogeneous in its amino acid sequence glycosylation can result in the production of $(36 \times 36)/2 = 648$ structurally unique IgG molecules or glycoforms; NB: the total number of combinations is divided by two because of the two-fold symmetry of the molecule. Analysis of monoclonal and polyclonal IgG demonstrates the presence of all the predicted oligosaccharide species, however, disialylated oligosaccharides may be absent or present at a very low level [6]. With the additional possibility of the presence of complex N-linked oligosaccharide in the Fab region, it is apparent that glycosylation is a post-translational modification that can introduce a very significant structural and, possibly, functional heterogeneity into the IgG molecule. The presence of additional glycosylation sites within the heavy chains of the other Ig isotypes means that the possible number of glycoforms may be orders of magnitude higher [7, 8].

The Fc glycosylation site is a conserved feature for all mammalian IgGs investigated and glycosylation occurs at a homologous position in human IgM, IgD and IgE, molecules, but not in IgA. Human IgM, IgA, IgE and IgD molecules bear additional N-linked oligosaccharide moieties attached to the constant domains of the heavy chains and IgA subclass 1 (IgA1) and IgD proteins also bear multiple 0-linked sugars in their extended hinge regions, attached to hydroxyl groups of serine and threonine residues. It has been estimated that $\sim 30\%$ of polyclonal IgG molecules also bear an oligosaccharide moiety within the Fab region. Since the sequences of the constant region of κ and λ light chains and the C_H1 domain of the heavy chain do not include a glycosylation sequon, the oligosaccharide of glycosylated Fab regions is due to attachment within either the V_L or V_H sequences. Analysis of the DNA sequences of 83 human germline V_H gene segments revealed five that encoded potential glycosylation sites, however, none of these sequons were observed in 37 V_H protein sequences – detailed analysis to determine whether the germline gene from which these proteins were derived did encode a glycosylation sequon was not attempted. Fifteen of the 37 protein sequences did have potential glycosylation sequons which, it would appear, have resulted from somatic mutation and antigen selection [7, 8]. It has been demonstrated that the structure and function of Fab oligosaccharide can depend on the site of attachment. Thus, monoclonal murine anti-dextran antibodies with a single oligosaccharide attachment site at residues 54, 58 or 60 in complementarity-determining region 2 (CDR2) were shown to have differing antigen-binding activities [9]. A monoclonal human polyreactive autoantibody, secreted by a heterohybridoma cell line was shown to be glycosylated on both the V_L and V_H regions; the V_H glycosylation site was at residue 75 in framework 3, adjacent to CDR3 [10].

Structural Consequences of IgG Glycosylation

Whilst glycosylation of the IgG/Fc is essential for optimal expression of effector activities mediated through FcγR and the C1 component of complement direct interaction of the oligosaccharide moiety with these effector ligands has not been demonstrated. Recently, it was reported, however, that in agalactosylated IgG the oligosaccharide moiety 'flips' out of the inter-C_H2 space and the terminal N-acetylglucosamine residues become available to bind and activate mannan-binding protein [11], and consequently the classical complement cascade. By contrast, these residues are not available to the lectin *Bandeireae simplicifolia* II in the native form of agalactosylated IgG but become so on denaturation. Resolution of structure for the oligosaccharide chains in x-ray crystallography demonstrates that it is not freely mobile and has definite conformation. From its attachment point at Asn297, it 'runs forward' towards the C_H2/C_H3 interface region and it is estimated that 82 non-covalent interactions between core sugar residues and the outer arm residues of the $\alpha[1 \rightarrow 6]$ arm may be possible [12]. Together with the sugars of the $\alpha[1 \rightarrow 3]$ arm, the oligosaccharide fills the available volume between the C_H2 domains. It may be anticipated, therefore, that whilst interactions with the Fx face of the protein impose structure on the oligosaccharide chain there is a reciprocal influence of the oligosaccharide on the protein structure.

The structural and functional consequences of Fc glycosylation can be assessed by comparison of glycosylated and aglycosylated forms of IgG. The latter can be generated by production in *Escherichia coli*, growing IgG producing cells in the presence of the glycosylation inhibitor tunicamycin or by protein engineering of the glycosylation sequon. It should be appreciated that IgG produced in *E. coli* or in the presence of tunicamycin will have an asparagine residue at 297 whilst site-directed mutagenesis can introduce any chosen amino acid residue; in the present case, alanine. A more subtle approach is to isolate homogeneous glycoforms for structural and functional studies or to replace individual or combinations of amino acid residues that make contacts with sugar residues. The latter approach may allow a detailed understanding of the oligosaccharide/protein interactions in this molecule, the 'rules' of template direction and its effect on the type of oligosaccharide attached and the generation of mutant molecules with new profiles of biologic function.

A small, localised protein structural change has been detected for aglycosylated human chimeric IgG3 and its Fc fragment by ^{1}H-NMR. Previous studies had allowed assignments for each of the five histidine residues, and their distribution through the Fc makes them suitable probes for detection of localised structural change. Such a change was reported for His268 which is in the vicinity of both the carbohydrate attachment site and the lower hinge binding site on IgG for Fc receptors [13]. A similar spectral difference was observed between a glycosylated IgG1 Fc fragment and the aglycosylated form produced as a recombinant protein in *E. coli* [14]. A structural difference between the lower hinge regions (residues 234–237) of glycosylated and aglycosylated IgG was inferred from the different papain cleavage profiles obtained for glycosylated and aglycosylated mouse IgG2b. Whilst a single cleavage point at residue 229 was observed for the glycosylated protein, the aglycosylated mutant was cleaved heterogeneously at residues 228, 234, and 235 [15].

Recent ^{13}C-NMR studies have provided direct evidence of differing structural dynamics for the lower hinge residues of glycosylated and aglycosylated mouse IgG2b [Kato, Lund and Jefferis, unpubl.]. Significant differences are revealed when thermodynamic parameters are determined from data obtained from differential scanning microcalorimetry of glycosylated and aglycosylated mouse IgG2b-Fc [Tischenko, Lund and Jefferis, unpubl.]. Differences are observed in both the C_H2 and C_H3 domains and the free energy of stabilisation of the C_H2 domain is decreased. An attempt to monitor structural differences between glycosylated and aglycosylated human IgG3 through altered epitope expression, employing a panel of > 30 mouse monoclonal anti-human Fcγ antibodies, did not detect any loss of expression or obvious reduction in affinity.

Functional Consequences of Asn297 Glycosylation

Since it has been consistently demonstrated that glycosylation is essential for optimal expression of FcγR- and C1-mediated effector functions, it may be anticipated that biological activity may vary between differing glycoforms. Most studies have compared differences between natural forms of IgG and their aglycosylated or agalactosylated counterparts; however, we have added the approach of generating mutant proteins in which residues reasoned to participate in oligosaccharide-protein interactions have been replaced.

A wide range of effector cells are activated by IgG/antigen immune complexes through interactions with cellular receptors for the Fc region of the gamma chain, FcγR. Three types of human Fc receptors (FcγRI, FcγRII, and FcγRIII) have been defined, by gene cloning and sequencing, that are differentially expressed on a variety of cell types; additionally FcγR may be induced or their expression up-regulated following cellular activation. The IgG isotype specificity of the FcγR suggests that recognition is correlated with primary amino acid sequence. An earlier prediction that the lower hinge residues 234–237 (-Leu-Leu-Gly-Gly-) in particular might correlate with FcγRI recognition appeared to be confirmed by protein engineering studies with the demonstration that replacement of any one of these residues in mouse/human chimeric IgG3 affected recognition by all three human FcγR [16, 17]. It was proposed, therefore, that the three Fc receptors are recognised by overlapping, non-identical ligand-binding sites. This appears rational for a family of receptors that are evolutionarily related and exhibit a high degree of sequence homology.

Other structural features are also determinants for recognition since replacement of Pro331 by serine, the amino acid residue present at this position in IgG4, reduces the binding affinity for IgG1 and IgG3 by an order of magnitude [18]; these residues are within 11 Å of the lower hinge. The FcγRII receptor is polymorphic and the allelic forms are designated as FcγRIIa-H131 and FcγRIIa-R131 to indicate that a histidine/arginine interchange at residue 131 is critical to recognition of human IgG2 molecules [19]. Thus, monocytes of homozygous H/H131 individuals were found to internalise IgG2-opsonized erythrocytes more efficiently than cells from R/R131 individuals. Since the lower hinge region of IgG2 molecules is radically different from that of IgG1 and IgG3, it is apparent that the recognition site for FcγRII depends on structure outside this region.

The amino acid replacement studies suggest that FcγR recognition is dependent on a precise molecular architecture and that subtle structural changes have a dramatic effect on biological function. This conclusion is further supported by the demonstration that aglycosylated human chimeric IgG3 has a reduced interaction with all three Fc receptors [16, 17, 20]. Whilst hapten-derivatised red blood cells could still be sensitised with this antibody to trigger superoxide production by U937 cells, stimulated with γ-interferon, higher levels of sensitisation were required compared to glycosylated IgG3 [20]. The aglycosylated IgG3 was not recognised by human FcγRII expressed on K562 and Daudi cells [21] and rosette formation mediated through FcγRIII, expressed on human NK (natural killer) cells, was reduced to 40% of that obtained for glycosylated IgG3, whereas antibody-dependent cellular cytotoxicity (ADCC) was essentially abolished [17]. Comparative studies of a glycosylated and an aglycosylated humanised anti-CD3 antibody suggest that the altered biological activities of aglycosylated IgG may be exploited for some in vivo applications. In the model investigated the glycosylated IgG was able to effect immune modulation and was immunogenic; probably due to its ability to activate T cells following interactions with appropriate FcγR-expressing cells. By contrast, the aglycosylated antibody was not immunomodulatory, was less immunogenic and had a longer half-life [22].

Attempts to evaluate the contribution of outer arm sugars to biologic function have concentrated mostly on glycoforms differing in galactose content. In a sustained investigation of the EBV-transformed lymphocytes secreting anti-D antibody, it has been shown that antibody with a high galactose content (>70% digalactosyl IgG) was more effective than antibody with a low galactose content (10% agalactosyl and 50% monogalactosyl IgG) in FcγRI- and FcγRIII-mediated cellular lysis (ADCC) [23]. An evaluation of the contribution of galactosylation to FcγRI recognition was made by comparison of the ability of a low galactose (<20% galactosylated) and a fully galactosylated form of an IgG4 Fc to inhibit superoxide generation through mouse/human chimeric IgG3, no difference was detectable in this system [24]. A minimal reduction in FcγR and a 2-fold reduction in C1q binding for agalactosyl IgG relative to the galactosylated form has been reported [25].

Considerable clinical experience has been gained with the humanised monoclonal antibody Campath-1H and its promise requires optimisation of control and efficiency of production. The product of rat YO, chinese hamster ovary (CHO) and mouse NSO cells has been evaluated for glycosylation and ADCC activity [26]. Interestingly, the rat cells were demonstrated to secrete IgG with relatively high levels of bisecting GlcNAc and to be the most active of the three products in ADCC, leading to the conclusion that this glycoform may have significant biologic advantage. The product of the NSO cells was reported to be underglycosylated. The final conclusion was that the cell type was a more important parameter than the culture conditions, at least for medium with and without added serum. It should be noted, however, that the method of culture used for each cell type was significantly different; the YO cells were grown in roller bottles, the NSO cells in shaking flasks and the CHO cells in hollow fibre bioreactors. In our experience [27], these differences in the method of culture could account, in large part, for the differences in glycoform profiles observed. The influence of outer arm sugars was evaluated for Campath-1H antibody following exposure to neuraminidase and β-galactosidase; removal of low levels of sialic acid had no effect on ADCC or complement-mediated lysis (CML), however, whilst removal of galactose was without effect on ADCC, it resulted in ~50% reduction in CML activity [28].

The essential requirement for protein/core-oligosaccharide interactions with a biantennary-type oligosaccharide is suggested from studies of a chimeric mouse-human IgG1 antibody produced in Lec-1 cells which are incapable of processing high mannose forms of oligosaccharide [29]. The antibody product having a high mannose oligosaccharide attached at Asn297 was incapable of complement-mediated hemolysis and deficient in C1q and FcγRI binding. In contrast a chimeric mouse-human IgG1 antibody produced in yeast cells, with presumed incorporation of high mannose forms of oligosaccharides at Asn297, maintained the ability to trigger ADCC through human FcγRIII [30]. The importance of C_H2 domain protein/core-oligosaccharide interactions in IgG is emphasised by the demonstration that recognition by Fcγ receptors can be modulated in mutant proteins in which core oligosaccharide contact residues have been replaced. Thus, replacement of Asp265, a contact residue for the primary GlcNAc residue of the core oligosaccharide, resulted in reduced recognition by human FcγRI and human FcγRII. By contrast, replacement of non-core contact residues Lys246, Asp249 by Ala and Glu258 by Asn was without effect on recognition for these receptors, a finding consistent with the view that the interactions with GlcNAc and Gal residues of the Manα(1→6) arm are not essential for maintenance of recognition by human FcγRI and FcγRII [24].

The biological half-life of a recombinant glycoprotein is a vital property determining in vivo efficacy and the economics of treatment. Studies of blood clearance of glycosylated and aglycosylated mouse/human chimeric IgG1 in mice demonstrated accelerated clearance for the aglycosylated form but with similar half-lives. Since the half-life of IgG1 in humans is ~23 days but measured as 5 days in this model it is difficult to draw a definitive conclusion. Catabolism of aglycosylated mouse IgG2b was evaluated in a rat model and shown to be cleared more rapidly than the glycosylated form and it was concluded that the increased catabolism occurred in the extravascular space [31]. The plasma half-lives and bioavailability of human anti-D antibodies secreted by Epstein-Barr virus (EBV)-transformed human B cells, cultured in hollow-fibre bio-reactors, have been evaluated in vivo in comparison with polyclonal anti-D isolated from immunised volunteers [32]. The half-lives of an IgG1 and an IgG3 monoclonal anti-D antibody were 22.2 and 10.2 days, respectively, compared to 15.6 days for polyclonal anti-D IgG. The half-life of polyclonal anti-D IgG was dependent on the proportions of IgG1 and IgG3 present in the preparation. Studies of mutant mouse IgG1 proteins have been interpreted to localise the site controlling catabolism to the inter C_H2/C_H3 region and to demonstrate modulation of the half-life [33].

Evaluation of a panel of 28 mutant mouse IgG2b proteins, each with a surface accessible amino acid replacement, for C1q binding and C1 activation correlated recognition with the presence of the wild-type residues lysine, glutamic acid and glutamic acid at 318, 320 and 322 [34]. A contrary result has been reported for mouse/human chimeric IgG1 antibody with the demonstration that replacement of glutamic acid 320 was without effect on CML, however, C1 activation was abrogated following amino acid replacements in the lower hinge region [35]. This is consistent with the observation that a Pro→Ser replacement at residue 331 in IgG1 and IgG3 results in a reduced capacity to trigger complement lysis [36]. One of the mutant proteins produced by Duncan and Winter [34] was Asn→Ala, 297 which results in the production of aglycosylated mouse IgG2b. This protein had a 3-fold reduced capacity to bind human C1q and a much reduced ability to trigger lysis of target cells with guinea pig complement through the classical complement cascade. Similarly, an aglycosylated mouse/human chimeric IgG1 was shown to retain some ability to trigger lysis of target cells by human complement but with a 7- to 8-fold higher antibody concentration requirement than for the glycosylated wild type IgG1 [37]. These data suggest a similarity in the molecular requirements for FcγR and C1 recognition and that glycosylation is essential for generation of a quaternary structure expressing these ligand binding sites.

The role of outer arm sugars in C1 mediated lysis has been investigated for galactosylated and agalactosylated IgG, produced following exposure to β-galactosidase, with an observed 2-fold higher activity for the galactosylated form [25]. Confirmation of the importance of correct glycosylation is provided by study of a human-mouse chimeric IgG1 molecule produced in yeast cells and anticipated to have high mannose type oligosaccharide attached at Asn297 [29]. The IgG1 product was unable to activate C1 to trigger human complement mediated lysis of targets whilst the same chimeric IgG1 construct expressed in rodent cells (Sp2/0) was effective. A direct role for the oligosaccharide moiety in activating the complement cascade is apparent for the lectin mannan-binding protein which can function as a surrogate C1 component. The specificity of mannan-binding protein is for mannose and N-acetylglucosamine residues, and it has been shown that it can access and bind to terminal N-acetylglucosamine residues exposed on agalactosyl IgG [11].

Much interest has been generated by the observation of a deficit in IgG galactosylation in patients with rheumatoid arthritis (RA) and some other inflammatory diseases, including tuberculosis and Crohn's disease [38]. Another feature of RA is the presence in the blood of rheumatoid factor (RF) autoantibodies having specificity for epitopes in the Fc region of IgG. Since RFs are, typically, of IgM or IgG isotype, the immune complexes formed have the potential to trigger effector functions through IgG-mediated pathways or a combination of IgG and IgM pathways. The chronic inflammatory reactions resulting are thought to contribute erosive damage in this disease. A dominant specificity of RFs is for an epitope localised to the area of contact and interaction between the C_H2 and C_H3 domains. This specificity overlaps with that of Staphylococcal protein A and the binding of a majority of RFs to IgG can be inhibited by Staphylococcal protein A [8, 39]. It has been speculated that terminal galactose or sialic acid residues on the α[1→6] arm of the oligosaccharide may be accessible to RFs and influence recognition and binding affinity, with a consequent effect on the nature and size of immune complex formed. A galactose residue on the α[1→6] arm is resolved on x-ray crystallography and possible non-covalent contacts identified. It has been argued, therefore, that this galactose residue occupies a lectin-like pocket that will be exposed in agalactosylated IgG and may contribute to altered IgG antigenicity, e.g. reactivity with RFs [38]. Alternatively, the mannose-binding protein provides a route by which agalactosylated IgG could trigger the inflammatory reactions seen in RA independently of RFs [11].

A study of the reactivity of 16 monoclonal RFs generated from synovial tissue lymphocytes with IgG of differing galactose content (18–86%) yielded ambiguous results. Five RFs reacted more avidly to IgG of low galactose content, 6 were not influenced by galactose content and one bound more avidly to IgG of high galactose content [40]. A comparison of the binding of polyclonal and monoclonal RFs to glycosylated and aglycosylated chimeric mouse/human IgG proteins of each of the subclasses detected no differences for IgG1, IgG2 and IgG4 proteins, however, RFs reactive with IgG3 proteins reacted more avidly with aglycosylated IgG3 [41]. In a companion study, some monoclonal RFs were found to bind aglycosylated IgG4 less well than glycosylated IgG4 (2- to 5-fold), suggesting that the carbohydrate moiety is important in establishing their binding epitope in the C_H2 domain [42]. An interesting difference between the latter two studies was that for one the source of monoclonal RF was serum of patients with Waldenströms macroglobulin-emia [41] and for the other EBV-transformed synovial tissue lymphocytes of RA patients. Given the parallelism between RF and Staphylococcal protein A binding to IgG, it is pertinent to note that there is only one report of a minimal effect of glycosylation on the binding of Staphylococcal A to IgG. An interesting recent study demonstrated isotype regulation mediated through the generation of auto-anti-isotype antibodies (RFs) during the course of an immune response to influenza virus. A series of RFs were established as monoclonal antibodies and demonstrated to effect immune deviation in vivo. One of these RFs, a monoclonal IgA RF, specific for mouse IgG2b bound the aglycosylated protein poorly [43].

On complexing with polyvalent antigen, IgM is able to initiate the classical complement cascade, following binding of C1q molecule to the C_H3 domain; the equivalent of the C_H2 domain of IgG. Amino acid replacements within glycosylation sequon 402–404 [44] of the C_H3 domain of mouse IgM results in a 3- to 25-fold decrease in the capacity to effect CML of target cells by guinea pig complement. This lowered activity could be due, at least in part, to an observed 4- to 8-fold reduction in assembly of the monomeric subunits into pentameric and hexameric IgM molecules. Replacement of residue 406 (Ser→Asn), analogous to core contact residue 301 in IgG, resulted in a 50-fold decrease in the capacity of mouse IgM to trigger lysis through guinea pig complement [45]. These data suggest that interactions between amino acid residues and core sugar residues of the oligosaccharide attached at Asn-402 of the IgM molecule may be important for the formation of the C1-binding and activation site.

Factors Influencing Glycosylation Hybridoma, and Recombinant Immunoglobulin Molecules

Regulatory authorities demand exhaustive testing of monoclonal antibodies that might be applied for in vivo diagnostic or therapeutic purposes. If approved, a similar demand for the demonstration of product consistency is made. The parameters that should be analysed, in vitro, include isotype, subclass, affinity, microheterogeneity, molecular weight, primary and secondary structure, structural integrity, specificity, glycosylation profile, biological potency. Subsequently, the product would be evaluated for pharmacological, toxicological, biodistribution and half-life in vivo [3]. Functional studies of recombinant human proteins have established that the form of the oligosaccharide moiety attached at a specific glycosylation site should be the same as that attached to the natural molecule. Regulatory authorities require authentic and consistent glycosylation of molecules that may be applied in vivo, therefore, animal cells are preferred to other systems for their production. The biotechnology industry has concentrated on development of production protocols employing CHO cells for all recombinant human glycoproteins demonstrating that its glycosylation machinery is catholic and that the polypeptide chain has a major influence on the type of oligosaccharide attached. However, CHO cells do not satisfy industrial economic requirements for the production of antibodies and so there has been a resurgence of interest in NSO cells that are derived from an antibody secreting plasmacytoma. A recent study also employed the rat plasmacytoma line YO and demonstrated its product to have a natural glycosylation profile that included the presence of bisecting GlcNAc residues [26]. For any given cell type, glycosylation of antibody products remains a variable dependent on numerous parameters that include, the method of cell culture, the supply of nutrients, removal of metabolic products, when the protein is harvested and a subtle influence of the polypeptide chain on outer arm sugar heterogeneity. A further concern is the possibility that mutant clones may arise during extensive and continuous culture with the emergence and overgrowth of a sub-clone secreting structurally and functionally aberrant molecules. The reality of this concern is demonstrated by the isolation of multiple sub-clones of CHO cells each of which expresses an altered profile of glycosyltransferases and consequently secretes glycoproteins with unique glycoform profiles [46].

Experience in the production of mouse/human chimeric antibodies in J558L cells demonstrated significant differences in galactosylation depending on whether it was produced in shallow culture, hollow fibre bioreactor or in vivo, as ascitic fluid. Of particular concern is the production of variable proportions of molecules bearing additional galactose residues in $\alpha[1\rightarrow3]$ linkage to normal galactose sugars. This results from the activity of an endogenous $\alpha[1\rightarrow3]$-galactosyltransferase. Gene expression for this enzyme is depressed in humans and higher primates with the result that it constitutes an immunogenic structure and it has been estimated that 1% of circulating human IgG is 'anti-Gal' antibody [47] and its presence can be readily demonstrated in an ELISA [48]. Although the CHO cell line expresses an $\alpha[1\rightarrow3]$ galactosyltransferase, there appears to be only one documented instance in which it has been demonstrated to be active in the addition of Gal $\alpha[1\rightarrow3]$ Gal [49]. We have had the experience of culturing multiple clones of transfected J558L whose antibody product had an essentially normal glycoform profile over several years; then, for reasons unknown to us, the antibody product included high mannose oligosaccharides and low site occupancy [Lund, Takahashi and Jefferis, unpubl. obs.]. A similar experience has been reported for human IgG1 and IgG2 antibodies produced by heterohybridomas for which variable proportions of high mannose containing antibody was obtained [50]. Glycosylation appeared to be dictated by the mouse plasmacytoma partner since N-glycolylneuraminic acid but no bisecting GlcNAc was added. Similarly, a humanised anti-CD18 antibody produced in NSO cells was shown to contain five oligomannoside-type structures in addition to the usual biantennary-type oligosaccharide moieties, no bisecting GlcNAc and no sialic acid [51]. The glycosylation status of human anti-D antibody produced by EBV-transformed lymphocytes grown at low density in static culture or high density in hollow fibre bioreactors also demonstrated high levels of galactosylation for antibody produced at low density and a relatively natural profile of glycoforms for antibody produced in the bioreactor [23]. Heterohybridomas secreting anti-D antibody have also been established; however, analysis showed that 12/16 such cell lines had incorporated Gal$\alpha[1\rightarrow3]$Gal epitopes into the antibody [52].

A further rodent/human difference is in the form of sialic acid utilised. Polyclonal human IgG has a terminal N-acetyl neuraminic acid sugar on $\sim25\%$ of oligosaccharides, by contrast the mouse utilises N-glycolyl neuraminic acid. Interestingly, chimeric mouse/human IgG3 produced in J558L cells was shown to be a mixture of molecules having one or the other derivative. This demonstrates that both transferases are available and that their utilisation is affected by subtle structural effects. A bisecting GlcNAc residue is present in $\sim10-20\%$ human polyclonal IgG but NSO and CHO cells lack the GlcNAc transferase III enzyme required for its addition. The extensive functional studies reported for recombinant molecules produced in CHO cells suggested that bisecting GlcNAc has little influence on biological activity, however, the presence of glycoforms with bisecting GlcNAc produced by rat YO cells has been held to account for its beneficial biological activity [26]. These findings point to the need for a productive cell line of human origin, however, none is available that has a high endogenous rate of protein synthesis. Many other vehicles for recombinant protein production are being appraised or under development. The early promise of *E. coli* has not been realised for glycoproteins since bacteria do not have a glycosylation machinery [14]. Experiences with insect cells differ. Thus whilst a mixture of high mannose and complex N-linked oligosaccharides was reported for recombinant human plasminogen, including a fully elongated biantennary form (28%) [53] the conclusion drawn from a study of the N-glycosylation of a virion protein was that insect cells were not capable of elongation with the addition of galactose and sialic acid [54]. It is evident that this system is very sensitive to culture conditions and the timing of infection with baculovirus. It is unlikely that other expression systems, such as transfected potatoes, tomatoes, are likely to allow production of glycoproteins that will be acceptable for therapeutic use; it should be remembered that glycosylation is only one of several post-translational events that are essential to the synthesis of a natural form of proteins and glycoproteins.

Whilst one might attempt to develop optimal growth conditions for basic scientific studies, it is likely that they would be too costly to translate into commercial production protocols. Ideally, one would aim for a system that mimics in vivo conditions (homeostasis!) as closely as possible with the maintenance of nutrient concentrations, oxygen tension, removal of metabolites. An unknown factor is the presence of essential growth factors (cytokines) in vivo. A major consideration for biotechnology companies is the overall cost of production and an important element in its determination is downstream processing. Isolation and purification is simplified by the use of defined media and there has been a sustained development of serum-free media with most companies adopting their own undisclosed formulation. Large-scale production facilities have employed air-lift fermenters of 10–12,000 litres capacity. There is a gradual scale-up with the growth of a 'charge' for the next fermenter to allow exponential growth. At the end stage, the cells exhaust the medium, die and protein is released following rupture of the cell wall.

Hollow-fibre bio-reactors have been used for research and intermediate scale production of glycoproteins, including antibodies to be used as in vivo therapeutics. This system does allow continuous exchange between the medium that the cells are suspended in and the 'external' circulating medium. However, the cells are not homogeneously dispersed throughout the cell compartment but grow in clumps of solid tissue with the result that mass transfer across such a tissue is inefficient and necrosis follows.

We have commented on the heterogeneity of glycosylation of IgG produced in vivo by healthy human adults and instanced altered galactosylation patterns in certain inflammatory diseases. It remains to be determined whether the IgG producing plasma cells are or are not abnormal per se or are developing and producing IgG in an abnormal environment. A fundamental question is whether this is a disease-specific phenomenon that has direct implications for cause and progression or an epiphenomenon that may be used to monitor disease activity and may have value as a prognostic indicator, as for α1-acid glycoprotein [55]. Analysis of mouse and human monoclonal IgGs has demonstrated that each clone exhibits a unique Fc glycosylation profile and, therefore, that the profile for polyclonal IgG is the sum of the many contributing clones. The human monoclonal IgG proteins have been isolated from sera of patients with the disease multiple myeloma. Analysis of a panel of IgG paraproteins with multiple examples of each subclass revealed a subtle template direction effect such that the apparent preference, in polyclonal IgG, for galactosylation of the $\alpha[1 \rightarrow 6]$ arm over the $\alpha[1 \rightarrow 3]$ arm was reversed for IgG2 proteins and for 2/3 IgG3 proteins [6]. In an extension of these studies, we have observed the $\alpha[1 \rightarrow 3]$ preference for a further five IgG3 paraproteins whilst the polyclonal IgG isolated from the same serum sample demonstrated the opposite preference [56]. A further observation is hypogalactosylation of both the polyclonal and the monoclonal IgG, relative to normal polyclonal IgG; however, this did not correlate with IL-6 levels in the same serum sample. It has been shown that IL-6 is a growth factor for plasma cells proliferating in the bone marrow in this disease and that it is reflected in increased IL-6 levels in the serum [57].

For proteins with multiple glycosylation sites, fidelity is observed for the type of oligosaccharide attached at each. Such template direction, excluding outer arm sugars, is exemplified for IgM, IgA, IgE and IgD molecules. Thus, for a mouse IgM secreting plasmacytoma, the oligosaccharide attached at Asn171 is a biantennary complex form, oligosaccharides at Asn332, 364 and 402 are triantennary and at Asn563 predominantly chitobiose ($Man_3GlcNac_2$) [58]. Similarly, in the human IgD molecule an oligomannose form is attached at residue 354 within the C_H2 domain, and complex forms at residues 445 and 496 within the C_H3 domain [59, 60]. While large-scale steric effects have been invoked in influencing accessibility of glycosylation sites to glycosylation enzymes, it is less widely appreciated that smaller-scale changes can also affect glycan synthesis. Repositioning of a carbohydrate attachment site within the Fab region of an antidextran antibody by two residues from Asn58 to Asn60 [9], resulted in the attachment of oligomannose forms in place of a complex form and was accompanied by ≥ 3-fold reduction in affinity for antigen. This study employed mouse hybridoma cells for antibody production and noted that the Fab oligosaccharide was more fully processed than the Fc moiety and that Galα[1→3]Gal was added to a significant proportion of Fab oligosaccharide. A more subtle influence of glycosylation status has been demonstrated for a human hybridoma that has a glycosylated λ-chain; variations in glucose availability were shown to determine the size of the oligosaccharide attached and the antigen-binding activity [61]. Alternatively, it has been shown that glycosylation sequons can be introduced into variable regions with consequent glycosylation that does not affect antigen binding and which can be used for conjugation of haptens [62].

The extensive interactions between the oligosaccharide and protein moieties suggest the possibility to modulate them by selected amino acid replacements with a consequent influence on biological activity. Such an effect has been observed on replacement of the core contact residue Asp265 by Ala (DA265), resulting in greatly increased levels of galactosylation but a diminution of FcγRI-mediated function; 50% more oligosaccharide chains with galactose residues on both the α[1→3] and α[1→6] arms for mouse/human chimeric IgG3 produced in CHO cells [63].

In conclusion, it is evident that post-translational glycosylation of proteins can have subtle and more far-reaching structural and functional consequences. These consequences will be particularly manifest for recombinant glycoproteins produced in vitro but intended for in vivo application. A second rapidly developing area of interest results from the observation of altered glycosylation states for specific proteins correlating with disease and/or disease activity. The field is set to expand as sensitive technologies for determining oligosaccharide structures and profiles are now commercially available.

References

1   Dwek R: Glycobiology – towards understanding the function of sugars. Biochem Soc Trans 1995; 23:1–25.
2   Jenkins N, Curling E: Glycosylation of recombinant proteins: Problems and prospects. Enzyme Microb Technol 1995;16:354–364.
3   Schaffner G, Haase M, Geiss S: Criteria for investigation of the product equivalence of monoclonal antibodies for therapeutic and in vivo diagnostic use in case of change in manufacturing process. Biologicals 1995;23:253–259.
4   Burton D, Woof JM: Antibody effector functions. Adv Immunol 1992;52:1–48.
5   Beale D, Feinstein A: Structure and function of the constant regions of immunoglobulins. Q Rev Biophys 1976;9:135–180.
6   Jefferis R, Lund J, Mizitani H, Nakagawa H, Kawazoe Y, Arata Y, Takahashi N: A comparative study of the N-linked oligosaccharide structures of human IgG subclass proteins. Biochem J 1990; 268:529–537.

7   Jefferis R: Glycosylation of antibody molecules: functional significance. Glycoconjug J 1993;10: 357–361.

8   Jefferis R, Lund J, Goodall M: Recognition sites on human IgG for Fcγ receptors: The role of glycosylation. Immunol Lett 1995;44:111–117.

9   Endo T, Wright A, Morrison SL, Kobata A: Glycosylation of the variable region of immunoglobulin G-site-specific maturation of the sugar chains. Mol Immunol 1995;32:934–940.

10   Leibiger H, Hansen A, Schoenherr G, Seifert M, Wustner D, Stigler R, Marx U: Glycosylation analysis of a polyreactive human monoclonal IgG antibody derived from a human-mouse heterohybridoma. Mol Immunol 1995;32:595–602.

11   Malhotra R, Wormald MR, Rudd PM, Fischer PB, Dwek RA, Sim RB: Agalactosyl IgG activates complement via mannose-binding protein. Nat Med 1995;1:237–243.

12   Padlan E: x-Ray diffraction studies of antibody constant regions, in Metzger H (ed): Fc Receptors and the Action of Antibodies. Washington, American Society for Microbiology, 1990, pp 12–30.

13   Lund J, Tanaka T, Takahashi N, Sarmay G, Arata Y, Jefferis R: A protein structural change in aglycosylated IgG3 correlates with loss of huFcRI and huFcRIII binding and/or activation. Mol Immunol 1990;27:1145–1153.

14   Matsuda H, Nakamura S, Ichikawa Y, Kozai K, Takano R, Nose M, Endo S, Nishimura Y, Arata Y: Proton nuclear magnetic resonance studies of the structure of the Fc fragment of human immunoglobulin G1: Comparisons of native and recombinant proteins. Mol Immunol 1990;27:571–579.

15   Hindley S, Gao Y, Nash P, Sautes C, Lund J, Goodall M, Jefferis R: The interaction of IgG with FcγRII: Involvement of the lower hinge binding site as probed with NMR. Biochem Soc Trans 1993;21:337S.

16   Lund J, Winter G, Jones PT, Pound JD, Tanaka T, Walker MR, Artymiuk PJ, Arata Y, Burton DR, Jefferis R, Woof JM: Human FcγRI and FcγRII interact with distinct but overlapping binding sites on the human IgG. J Immunol 1991;147:2657–2662.

17   Sarmay G, Lund J, Rozsnyay Z, Gergely J, Jefferis R: Mapping and comparison of the interaction sites on the Fc region of IgG responsible for triggering antibody dependent cellular cytotoxicity (ADCC) through different types of human Fc receptor. Mol Immunol 1992;29:633–639.

18   Canfield SM, Morrison SL: The binding affinity of human IgG for its high affinity Fc receptor is determined by multiple amino acids in the $C_{H}2$ domain and is modulated by the hinge region. J Exp Med 1991;173:1483–1491.

19   Warmerdam PAM, van de Winkle JGJ, Gosseli EJ, Capel PJA: A single amino acid in the second domain of the human Fcγ receptor II is critical for human IgG2 binding. J Immunol 1991;172: 19–25.

20   Pound JD, Lund J, Jefferis R: Aglycosylated chimeric human IgG3 can trigger the human phagocyte respiratory burst. Mol Immunol 1993;30:233–241.

21   Walker MR, Lund J, Thompson KM, Jefferis R: Aglycosylation of human IgG1 and IgG3 monoclonal antibodies can eliminate recognition by human cells expressing FcγRI and FcγRII receptors. Biochem J 1989;259:347–353.

22   Routledge EG, Falconer ME, Pope H, Lloyd IS, Waldmann H: The effect of aglycosylation on the immunogenicity of a humanized therapeutic CD3 monoclonal antibody. Transplantation 1995;60: 847–853.

23   Kumpel BM, Rademacher TW, Rook GAW, Williams PJ, Wilson IBH: Galactosylation of human IgG monoclonal anti-D produced by EBV-transformed B-lymphoblastoid cell lines is dependent on culture method and affects Fc receptor-mediated functional activity. Hum Antibody Hybridomas 1994;5:143–151.

24   Lund J, Takahashi N, Pound JD, Goodall M, Nakagawa H, Jefferis R: Oligosaccharide-protein interactions in IgG can modulate recognition by Fcγ receptors. FASEB J 1995;9:115–119.

25   Tsuchiya N, Endo T, Matsuta K, Yoshinoya S, Aikawa T, Kosuge E, Takeuchi F, Miyamoto T, Kobata A: Effects of galactose depletion from oligosaccharide chains on immunological activities of human IgG. J Rheumatol 1989;16:285–290.

26   Lifely M, Hale C, Boyce S, Keen MJ, Phillips J: Glycosylation and biological activities of Campath-1H expressed in different cell lines and grown under different culture conditions. Glycobiology 1995; 5:813–822.

27 Lund J, Takahashi N, Nakagawa H, Goodall M, Bentley T, Hindley SA, Tyler R, Jefferis R: Control of IgG/Fc glycosylation: A comparison of oligosaccharides from chimeric human/mouse and mouse subclass immunoglobulin Gs. Mol Immunol 1993;30:741–748.

28 Boyd PN, Lines AC, Patel AK: The effect of the removal of sialic acid, galactose and total carbohydrate on the functional activity of Campath-1H. Mol Immunol 1995;32:1311–1318.

29 Wright A, Morrison S: Effect of altered C_H2 associated carbohydrate structure on the functional properties and in vivo fate of human immunoglobulin 1. J Exp Med 1994;180:1087–1096.

30 Horwitz AH, Chang CP, Better M, Hellstrom KE: Secretion of functional antibody and Fab fragment from yeast cells. Proc Natl Acad Sci USA 1988;85:8678–8682.

31 Wawrzynczak EJ, Cumber AJ, Parnell GD, Jones PT, Winter G: Blood clearance in the rat of a recombinant mouse monoclonal antibody lacking the N-linked oligosaccharide side chains of the C_H2 domains. Mol Immunol 1992;219:213–220.

32 Kim JK, Tsen MF, Ghetie V, Ward ES: Evidence that the hinge region plays a role in maintaining serum levels of the murine IgG1 molecule. Mol Immunol 1995;32:467–475.

33 Goodrick J, Kumpel B, Pamphilon D, Fraser I, Chapman G, Dawes B, Anstee D: Plasma half-lives and bioavailability of human monoclonal Rh D antibodies BRAD-3 and BRAD-5 following intramuscular injection into RhD-negative volunteers. Clin Exp Immunol 1994;98:17–20.

34 Duncan AR, Winter G: The binding site for C1q on IgG. Nature 1988;332:738–740.

35 Morgan A, Jones ND, Nesbitt AM, Chaplin L, Bodmer MW, Emtage JS: The N-terminal end of the C_H2 domain of chimeric human IgG1 anti-HLA-DR is necessary for C1q, FcγRI and FcγRII binding. Immunology 1995;86:319–324.

36 Tao MH, Smith RIF, Morrison SL: Structural features of human immunoglobulin G that determine isotype-specific differences in complement activation. J Exp Med 1993;178:661–667.

37 Dorai H, Mueller BM, Reisfeld RA, Gillies SD: Aglycosylated chimeric mouse/human IgG1 antibody retains some effector function. Hybridoma 1991;10:211–217.

38 Parekh RB, Dwek RA, Sutton JB, Fernandes DL, Leung A, Stanworth D, Mizuochi T, Taniguchi T, Matsuta K, Takeuchi F, Nagano Y, Miyamoto T, Kobata A: Association of rheumatoid arthritis and primary osteoarthritis with changes in the glycosylation pattern of total serum IgG. Nature 1985;316:452–457.

39 Jefferis R, Nik Jaafar MIB, Steinitz M: Immunogenic and antigenic epitopes of immunoglobulins. VIII. A human monoclonal rheumatoid factor having specificity for a discontinuous epitope determined by histidine/arginine interchange as residue 435 of immunoglobulin G. Immunol Lett 1984;7:191–194.

40 Soltys AS, Hay FC, Bond A, Axford AS, Jones MG, Randen I, Thompson KM, Natvig JB: The binding of synovial tissue derived monoclonal human immunoglobulin M rheumatoid factors to immunoglobulin G preparations of differing galactose content. Scand J Immunol 1994;40:135–143.

41 Artandi SE, Canfield SM, Tao MH, Calame KL, Morrison SL, Bonagura VR: Molecular analysis of IgM rheumatoid factor binding to chimeric IgG. J Immunol 1991;146:603–610.

42 Bonagura VR, Artandi SE, Davidson A, Randen I, Agostino N, Thompson K, Natvig J, Morrison SL: Mapping studies reveal unique epitopes on IgG recognised by rheumatoid arthritis-derived monoclonal rheumatoid factors. J Immunol 1993;151:3840–3852.

43 Rajnavolgyi E, Fazekas GY, Lund J, Daeron M, Teillaud J, Jefferis R, Fridman W, Gergely J: Activation of effector functions by immune complexes of mouse IgG2a with isotype specific autoantibodies. Immunology 1995;85:645–654.

44 Muraoka S, Shulman MJ: Structural requirements for IgM assembly and cytolytic activity. Effects of mutations in the oligosaccharide acceptor site at Asn-402. J Immunol 1989;142:695–701.

45 Shulman MJ, Pennel N, Collins C, Hozumi N: Activation of complement by immunoglobulin M is impaired by the substitution serine-406→asparagine in the immunoglobulin μ heavy chain. Proc Natl Acad Sci USA 1986;863:7678–7682.

46 Stanley P, Ioffe E: Glycosyltransferase mutants – key to new insights in glycobiology. FASEB J 1995;9:1436–1444.

47 Galili U, Shohet SB, Kobrin E, Stults CLM, Macher BA: Man, apes and old world monkeys differ from other mammals in the expression of α-galactosyl epitopes on nucleated cells. J Biol Chem 1988;263:17755–17762.

48   Borraebeck CAK, Malmbor AC, Ohlin M: Does endogenous glycosylation prevent the use of mouse monoclonal antibodies as cancer therapeutics? Immunol Today 1993;14:477–479.

49   Ashford DA, Alaf CD, Gamble VM: Site-specific glycosylation of rat and human soluble CD4 variants expressed in CHO cells. J Biol Chem 1993;268:3260–3267.

50   Tandai M, Endo T, Sasaki S, Masuho Y, Kochibe N, Kobata A: Structural study of the sugar moieties of monoclonal antibodies secreted by human-mouse hybridoma. Arch Biochem Biophys 1991;291:339–348.

51   Yu Ip C, Miller WJ, Silberklang M, Mark GE, Ellis RW, Huang L, Glusha J, Van Halbeek H, Zhu J, Alhadeff JA: Structural characterisation of the N-glycans of a humanized anti-CD18 murine immunoglobulin. G Arch Biochem Biophys 1994;308:387–399.

52   Montano RM, Romano EL: Human monoclonal anti-Rh antibodies produced by human-mouse heterohybridomas express the gal α(1–3) Gal epitope. Hum Antibody Hybridomas 1995;5:152–156.

53   Davidson DJ, Castellino FJ: Structures of the Asn-289 linked oligosaccharide assembled on human plasminogen expressed in *Mamestra brassicae* cell line (IZD.MB503). Biochemistry 1991;30:6689–6696.

54   Jarvis DL, Finn EE: Biochemical analysis of the N-glycosylation pathway in baculovirus-infected insect cells. Virology 1995;212:500–511.

55   van Dijk W, Turner GA, Mackiewicz A: Changes in glycosylation of acute-phase proteins in health and disease: Occurrence, regulation and function. Glycosyl Dis 1994;1:5–14.

56   Farooq M, Takahashi N, Arrol H, Drayson M, Jefferis R: Glycosylation of IgG antibody molecules in multiple myeloma. 1st Electronic Glycoscience Conference – proceedings, in press.

57   Nakao H, Nishikawa A, Nishiura T, Kanayama Y, Tarui S, Taniguchi N: Hypogalactosylation of immunoglobulin G sugar chains and elevated serum Il-6 in Castleman's disease. Clin Chim Acta 1991;197:221–228.

58   Anderson DR, Atkinson PH, Grimes WJ: Major carbohydrate structures at five glycosylation sites on murine IgM determined by high resolution 1H-NMR spectroscopy. Arch Biochem Biophys 1985;13:605–618.

59   Mellis SJ, Baenziger JU: Structures of the oligosaccharides present at the 3 asparagine-linked glycosylation sites of human IgD. J Biol Chem 1983;258:11546–11555.

60   Mellis SJ, Baenziger JU: Structures of the O-glycosidically linked oligosaccharides of human IgD. J Biol Chem 1983;258:11557–11563.

61   Tachibana H, Taniguchi K, Ushio Y, Teruya K, Osada K, Murakami H: Changes of monosaccharide availability of human hybridomas lead to alteration of biological properties of human monoclonal antibody. Cytotechnology 1994;16:151–157.

62   Leung SO, Losman LMJ, Govindan SV, Griffiths GL, Goldenberg DM, Hansen HJ: Engineering a unique glycosylation site for site-specific conjugation of haptens to antibody fragments. J Immunol 1995;154:5919–5926.

63   Lund J, Takahashi N, Pound JD, Goodall M, Jefferis R: Multiple interactions of IgG with its core oligosaccharide can modulate recognition by complement and human FcγRI, and influence the synthesis of its oligosaccharide chains. J Immunol 1996, in press.

Prof. Roy Jefferis, Department of Immunology, The Medical School,
University of Birmingham, Birmingham B15 2TT (UK)

Capra JD (ed): Antibody Engineering.
Chem Immunol. Basel, Karger, 1997, vol 65, pp 129–158

······················

Engineering Novel Antibody Molecules

Martha G. Sensel, M. Josefina Coloma, Eric T. Harvill, Seung-Uon Shin,
Richard I.F. Smith, Sherie L. Morrison

Department of Microbiology and Molecular Genetics and the Molecular Biology
Institute, University of California, Los Angeles, Calif., USA

Antibody molecules exhibit multiple functions that make them key components of the mammalian immune system, and their exquisite specificity offers
great potential for therapeutic applications. The tetrameric structure of the
molecule, which consists of two identical light chains and two identical heavy
chains joined by disulfide linkages, is integral to its function (fig. 1). The amino
acid sequence of the molecule dictates its structure: the light chains and heavy
chains fold into two and four or five functional domains, respectively, with
each functional domain consisting of seven anti-parallel β-strands forming a
β-barrel. The N-terminal domains from both the heavy and light chains are
called variable regions and contain the binding site which allows specific
recognition of antigens. The remaining domains, referred to as constant regions, interact with the effector arms of both the humoral and cellular immune
system. The ability of antibodies to both recognize antigens and stimulate the
immune effector systems is now being exploited in novel ways for treatment
of various diseases.

Advances in genetic engineering and expression systems have led to rapid
progress in the development of immunoglobulins with defined or novel functional properties. Moreover, the natural organization of the immunoglobulin
gene, which consists of separate exons encoding separate domains, has facilitated its manipulation (fig. 1). Functional domains that carry antigen-binding
activities (Fab, Fv) or effector functions (Fc) can be exchanged readily between
different antibodies (fig. 1), and novel nonimmunoglobulin proteins can be
added to either the variable or constant domains. Indeed, antibody molecules
have proven to be surprisingly amenable to such modifications. Production of
these novel antibodies has been accomplished via a number of expression

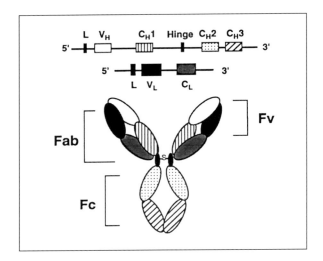

Fig. 1. Protein and genetic structure of an antibody molecule. For both heavy- and light-chain genes, each exon encodes a separate protein domain. Two heavy (H) and two light (L) chains comprise the antibody. Each chain consists of constant (C) and variable (V) regions. The heavy and light chains as well as the two heavy chains are held together by disulfide bonds. The variable region and the first constant domain of the heavy chain together with the light chain make up the Fab (fragment, antigen binding). The second and third constant domains of the heavy chain make up the Fc (fragment, crystallizable), and the heavy- and light-chain variable domains together comprise the Fv (fragment, variable domains).

systems, including bacteria, yeast, plants, baculovirus and mammalian cells. However, expression of intact, fully functional antibodies has been achieved most successfully in mammalian cells, which possess the mechanisms required for correct immunoglobulin assembly, posttranslational modification, and secretion. Posttranslational modifications can influence both the biological properties and effector functions of the antibody, which are important considerations when the antibody is intended for diagnostic and therapeutic use. Finally, to reduce immunogenicity, protein engineering has been used to convert murine antibodies both to chimeric antibodies containing mouse variable regions and human constant regions and to 'humanized' antibodies that contain only murine complementarity-determining regions.

The rationale for creating these various 'engineered antibodies' is multifaceted. Antibody-ligand fusion proteins have been attempted by numerous investigators since in theory they should exhibit dual specificities: both the antigen-combining regions and the ligand should bind to their respective binding partners. If both partners are contained within the same target, for example an antigen and a growth factor (GF) receptor on a tumor cell, enhanced binding specificity should result. If the antibody so engineered retains its effector functions, an immune response against the target may be achieved. In addition, domain exchange between isotypes and site-directed mutagenesis have been used to gain insight into structure-function relationships in the antibody molecule. Studies of these numerous engineered antibodies have proven to be fruitful not only in terms of development of potential therapeutic drugs, but also in terms of enhancing our basic understanding of antibody function.

Thus, current techniques have allowed production of a variety of antibodies with novel properties. Herein we will focus on a number of molecules produced in our laboratory. After a general introduction to antibody fusion proteins, we will discuss production and characterization of a set of antibodies containing either rat insulin-like growth factor 1 (IGF-1), human IGF-2, or transferrin (Tf) at their C-termini. Next we will describe an antibody fusion protein that consists of an IgG engineered to contain the cytokine IL-2 at its C-terminus. Finally, we will describe a family of IgGs engineered to form polymeric IgM-like molecules.

Antibody Fusion Proteins

Production of a fusion protein between nonimmunoglobulin and immunoglobulin molecules can be achieved in several different ways. One approach is to substitute the nonimmunoglobulin sequences for the antibody variable region. This method should result in a protein that has acquired antibody-associated properties such as effector functions or improved pharmacokinetics. A number of molecules engineered in this manner have been called 'immunoadhesins' because they contain a cell surface-binding molecule linked to the immunoglobulin Fc effector domain [1]. For example, immunoadhesins containing CD4, the T cell surface protein recognized by HIV, have been created [2, 3]. The therapeutic actions of immunoadhesins are in part conferred by the mAb moiety of the fusion protein, and therefore are depending on the mAb isotype used [2–5]. For example, an IL-2-IgG1 fusion protein was able to effect complement-mediated lysis of IL-2 receptor-bearing cells [6].

An alternative approach for engineering antibody fusions is to substitute or join nonimmunoglobulin sequences with the antibody constant region (fig. 2). Molecules engineered in this way retain the binding specificity of the antibody and, depending on the position of the substitution, may retain antibody-related effector functions and biological properties. Thus, with this approach, an antibody combining specificity can be used to provide specific delivery of an associated biological activity. For example, an antibody fusion combining high-affinity antifibrin binding specificity with tissue plasminogen

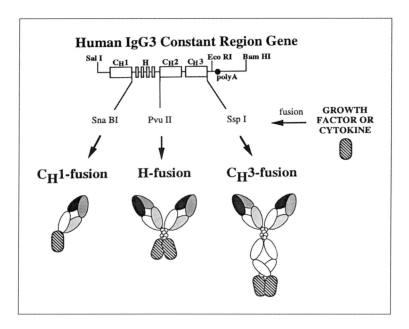

Fig. 2. Schematic diagram of proteins produced using IgG3-GF/cytokine cassette vectors. Site-directed mutagenesis was used to generate unique blunt end restriction enzyme sites in the human IgG3 heavy-chain gene at the 3' end of the C_H1 exon, immediately after the hinge at the 5' end of the C_H2 exon, and at the 3' end of the C_H3 exon (open boxes represent human IgG3 exons). An *EcoRI* site was also introduced 3' of the IgG3 gene to provide a 3' cloning site for the GF or cytokine and a poly A addition signal (closed circle). The IgG3-IGF fusion heavy-chain genes were joined to a mouse variable region (not shown) and expressed with the specific light chain. The resulting IgG3-GF fusion proteins are depicted diagrammatically.

activator resulted in a thrombolytic agent that is more specific and more potent than tissue plasminogen activator alone [7]. Fusion of tumor necrosis factor to an antitransferrin receptor (TfR) antibody resulted in a fusion protein with tumor necrosis factor cytotoxic activity towards cell lines expressing the TfR [8]. Similarly, antiganglioside GD2 specific IgG1 proteins containing human lymphotoxin joined after C_H2 or C_H3 were cytolytic for cells sensitive to lymphotoxin [9].

As described below, we have chosen this latter approach for engineering a number of monoclonal antibody (mAb) fusion proteins. It is possible that for different therapeutic uses, one would want to employ an mAb with either enhanced or diminished effector functions. Thus our goal has been to create a series of fusion proteins that will retain the activity of the nonimmunoglobulin partner and may or may not exhibit mAb-associated properties. In studies undertaken in our laboratory, cloning cassettes were constructed such that proteins could be joined at various positions to the constant regions of mouse-human chimeric IgGs. All of the fusion proteins to be discussed herein consist of a human IgG3 molecule combined with the ligand for a cell surface receptor. Human IgG3, unlike other IgG isotypes, contains an extended hinge region of 62 amino acids that should facilitate simultaneous antigen and receptor binding by providing spacing and flexibility. Site-directed mutagenesis of human IgG3 was used to generate unique restriction at the 3' end of the C_H1 exon, at the 5' end of the C_H2 exon (immediately after the hinge), and at the 3' end of the C_H3 exon (fig. 2). In most cases, PCR was used to generate clones encoding fully processed forms (lacking any sequences which are removed posttranslationally) of the non-immunoglobulin fusion partner. The variable region can be of any desired combining specificity. In all cases, the heavy-chain fusion genes are expressed with the light chain of the corresponding specificity in an appropriate recipient, such as a nonproducing myeloma cell line.

Antibody Fusion Proteins for Receptor-Mediated Targeting of the Brain

We first turn our attention to a group of mAb GF fusion proteins designed to cross the brain capillary endothelial wall that constitutes the blood-brain barrier (BBB). Drug delivery to the brain is limited by poor transport of water-soluble molecules through this barrier. Transport of nutrients and GF from the blood to the brain is accomplished via specific receptors on the BBB. For example, the BBB contains receptors for molecules such as insulin [10], Tf [11], and IGF-1, IGF-2 [12, 13]. These receptors provide potential vehicles for transport into the brain (fig. 3). Indeed, as described below, both anti-GF receptor mAbs and mAb-GF fusion proteins have been investigated for BBB drug delivery.

Although immunoglobulin molecules normally are excluded from the brain [14], an antibody to theTfR has been shown to be transported into the brain parenchyma and to be effective as a drug delivery vehicle [15]. Furthermore, this antibody was demonstrated to be an effective vehicle for the delivery of chemically conjugated nerve growth factor [16]. The mAb-nerve growth factor conjugate which retained nerve growth factor activity showed a pattern of brain localization similar to that observed with the unconjugated anti-Tf antibody, and intravenous injection of the conjugate prevented loss of striatal choline acetyltransferase-immunoreactive neurons in a rat model of Huntington disease [17].

An alternative approach for mAb-mediated targeting of GF receptor-bearing cells in the BBB is to use the GF itself as the targeting moiety. As

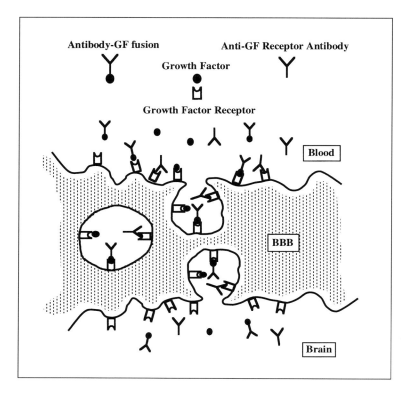

Fig. 3. Transport of proteins across the BBB via GF receptors. Anti-receptor mAb and mAb-GF fusion proteins should cross the BBB by binding to the GF receptor on the blood capillary endothelium. Vesicular transport across the BBB delivers the protein to the brain.

described above, the approach taken in our laboratory has been to engineer a family of vectors that contain GF genes fused to the mAb constant region domains. This system allows production of molecules that will retain their ability to bind both antigens and GF receptors.

In initial studies, a fusion protein consisting of IGF-1 joined immediately after the hinge of anti-dansyl (αDNS) IgG3 was produced [18]. This fusion protein retained its ability to bind both antigen and the IGF-1 receptor and was able to stimulate amino acid and sugar uptake by IGF-1 receptor-bearing cells. Experiments were then undertaken to expand these studies by creating a family of multifunctional antibody molecules capable of targeting different specific receptors through the attached GF [19, 20]. Because their receptors are abundant in the BBB, the following GFs were chosen for use: (1) rat IGF-1 (70 amino acids); (2) human IGF-2 (67 amino acids); (3) Tf (180 kD). As described above, in each case the GF was joined to either the C-terminus of C_H1 (C_H1-GF), the N-terminus of C_H2 (immediately after the hinge; H-GF), or the C-terminus of C_H3 (C_H3-GF) of IgG3. The proteins C_H1-GF, H-GF, and C_H3-GF were expected to exist as Fab, F(ab)$'_2$ and full-length IgG3, respectively. The vectors for these various constructs were stably transfected into nonimmunoglobulin-producing myeloma cells and the secreted proteins were purified from culture supernatants by affinity chromatography. Reducing and nonreducing SDS-PAGE analysis confirmed that correctly sized fusion proteins were produced: each heavy chain migrated with the molecular weight expected for mAb plus GF, and each fusion molecule was assembled into the predicted form (FAb, F(ab)$'_2$, or complete tetramer).

All of the mAb-GF fusion proteins retain their ability to bind the Ag DNS. This ability to bind antigen is an extremely important feature of these proteins and could potentially be utilized for secondary targeting with the brain, or in the case of antidrug antibodies, for delivery of the therapeutic agent. Although initial experiments used an antihapten antibody, other variable regions with different specificities can be substituted to generate fusion proteins with unique binding properties.

Binding of Fusion Proteins to Their GF Receptors

In order for these mAb-GF fusion proteins to be useful therapeutically, it is imperative that they retain the ability to bind to their respective GF receptors. Therefore, competition binding assays were employed to compare relative affinities of the fusion proteins with the native growth factors. For these assays, both the unlabeled fusion protein and IgG3 were used as competitors for ^{125}I-labeled GF binding to receptor-bearing cells. Results from these assays indicated that both H-IGF-1 and C_H3-IGF-1 bound to the IGF-1 receptor, albeit with reduced affinity compared to IGF-1. Concentrations required for 50% inhibition (IC$_{50}$) for both H-IGF-1 and C_H3-IGF-1 (approximately 1×10^{-6} M) were approximately 1,000-fold higher than that for IGF-1. The C_H1-IGF-1 Fab molecule was significantly impaired in its binding ability, achieving only 30% inhibition at the highest concentration (1.53×10^{-6} M) tested. Wild-type IgG3 showed no inhibition of ^{125}I-IGF-1 binding at any concentration. Therefore, competition by the fusion proteins occurred as a consequence of the presence of the IGF-1 moiety in each molecule.

In similar assays performed with ^{125}I-IGF-2, the IC$_{50}$s were approximately 10- to 50-fold higher for the three IGF-2 fusion proteins ($3–9 \times 10^{-8}$ M) than for recombinant IGF-2 (2×10^{-9} M). Surprisingly, inhibition was also observed for wild type IgG3, but the IC$_{50}$ (2×10^{-6} M) was approximately 100-fold higher than that for the fusion proteins. These results indicate that the IGF-2 moiety in the fusion protein enables it to specifically bind to the IGF-

2 receptor, albeit with decreased affinity compared to IGF-2. The differential ability of IgG3 to compete with ^{125}I-IGF-2 and ^{125}I-IGF-1 binding is likely due to the fact that IGF-1 and IGF-2 bind to different cell surface receptors: IGF-1 binds to an insulin-like receptor, whereas IGF-2 binds to the mannose-6-phosphate receptor.

Competition by each of the IgG3-Tf fusion proteins with ^{125}I-labeled Tf for binding to the TfR also was determined. The IC_{50} of competitor IgG3-H-Tf (3.1 nM) and IgG3-C_H3-Tf (2.6 nM) were slightly higher, but similar to that of Tf (1.2 nM), but the IC_{50} for IgG-C_H1-Tf (19.3 nM) was 15-fold higher than that for Tf. These differences may be attributable to the fact that the two Tf molecules in H-Tf and C_H3-Tf may facilitate binding to the dimeric TfR, whereas the single Tf molecule in C_H1-Tf can bind only to one Tf-binding site.

Wild-type IgG3 showed no inhibition of ^{125}I-Tf binding, even at high concentrations (6×10^{-6} M). Taken together, these results indicate that all of the IgG3-GF fusion proteins have functional GF moieties. Importantly, a number of the fusion proteins exhibit similar affinities for their GF receptor when compared to the native GF.

Effector Functions of the IgG3-C_H3-GF Fusion Proteins

The studies conducted above demonstrated that the GF moiety of the fusion protein is functional. Similarly, it is important to determine whether or not the mAb moiety retains mAb-associated functions. Isolation and purification of the fusion proteins using the Ag DNS demonstrated the functionality of the variable regions. Effector functions, such as complement (C′) activation [21–24] and Fc receptor binding [25–27], are associated with the C_H2 of the IgG molecule. Fusion proteins containing the GF joined to C_H3, but not C_H1 or hinge would be expected to retain these functions. Therefore, we examined the ability of the various IgG3-C_H3-GF fusion proteins both to activate C′ and to bind FcγRI on U937 cells.

C′-mediated lysis of Ag-coated, ^{51}Cr-loaded sheep red blood cells (SRBC) was compared for each of the fusion proteins and native IgG3. The ability of the IgG3-C_H3-IGF-1 fusion protein to activate C′ was reduced compared to wild type IgG3. Surprisingly, the ability of IgG3-C_H3-IGF-2 to activate C′ was increased approximately 50-fold compared to IgG3. In order to verify that this increased activity was due to Ag-specific activation of the classical pathway of C′ activation, control assays were performed using either uncoated SRBC or buffer containing EGTA, which chelates Ca^{2+} ions that are required for classical pathway activation. No activation of C′ by IgG3-C_H3-IGF-2 was observed using either of these conditions. In contrast, to the IgG3-IGF fusion proteins, the IgG3-C_H3-Tf fusion protein was unable to activate complement-mediated lysis of Ag-coated SRBC. However, in a C′ consumption assay, which involves preincubation of mAb, C', and soluble Ag prior to the addition of SRBC, IgG3-C_H3-Tf was active, albeit less so than wild-type IgG3.

Lysis of target cells represents the final step in the C' pathway, but differences in C'-mediated lysis may be due to differences in individual steps of the cascade. Thus, we employed an immunoassay system to determine whether the fusion proteins differed in C1q binding, the first step of the C' cascade. Results of the C1q-binding assay showed that both the IgG3-C_H3-IGF-1, IgG3-C_H3-IGF-2 and IgG3-C_H3-Tf were similar to wild-type IgG3. Thus, despite similarities in ability to bind C1q, the C_H3-IGF-1 and IgG3-C_H3-Tf fusion proteins exhibit impairment, whereas the C_H3-IGF-2 fusion proteins exhibit enhancement of their ability to induce C'-mediated lysis. These data indicate that fusion of the GF moiety to IgG3 affects subsequent steps in the C' cascade.

Type I IgG-specific Fc receptors (FcγRI) are expressed on a variety of lymphoid cell types and are important for host defense. We assessed binding of the IgG3-C_H3-GF fusion proteins to FcγRI expressed by interferon-γ-treated U937 cells by determining their ability to compete with ^{125}I-labeled IgG3. These assays yielded IC_{50} values of 7 nM for IgG3-C_H3-IGF-1, 5 nM for IgG3-C_H3-IGF-2, and 3.5 nM for IgG3-C_H3-Tf, which are similar to the value of 10 nM obtained for IgG3. Therefore, the IgG3-C_H3-GF fusion proteins resemble IgG3 in their ability to bind FcγRI, indicating that the GF moieties did not affect accessibility of the hinge proximal FcγRI binding site [25–27]. It is interesting to note that although both the C1q-binding site and FcγRI-binding sites are located within the IgG C_H2 domain, the IGF-1 and IGF-2 moieties had significantly different effects on complement-mediated hemolysis, but similar effects on FcγRI binding.

In vivo Targeting of the GF Fusion Proteins to the Brain

The studies described above demonstrated that the IgG3-GF fusion proteins retained their mAb-associated and GF-associated activities. We now examine the abilities of these proteins to carry out their intended function: targeting to the brain in vivo. For these experiments, the IgG3-GF fusion proteins were labeled with ^{3}H-succinimidyl propionate and injected into the tail vein of female Sprague-Dawley rats. The animals were sacrificed and the brains removed at various times after injection. The brain tissue homogenates were depleted of capillaries by density gradient centrifugation and the radioactivity in the brain parenchyma (postcapillary supernatant) and the brain capillary pellet were determined. Increased localization of radioactivity to the brain parenchyma over time indicates that material has crossed the BBB and has been taken up into brain. In all cases, ^{14}C-labeled IgG3 was used as an internal control to correct for blood contamination.

First we discuss the results of studies conducted with the IgG3-IGF-1 fusion proteins. Twenty-four hours following injection, similar amounts of IgG3-C_H1-IGF-1 (0.15% ID), IgG3-H-IGF-1 (0.15% ID), and IgG3-C_H3-IGF-1 (0.10% ID) were found in the brain parenchyma. The kinetics of uptake, however, depended on the construct. IgG3-C_H3-IGF-1 is taken up most rapidly, reaching peak parenchymal levels 1 h after injection, with none of the protein associated with the capillary fraction. IgG3-C_H1-IGF-1 is taken up more rapidly than IgG3-H-IGF-1, but the kinetics become similar by 4 h after injection. In contrast to IgG3-C_H3-IGF-1, significant amounts of both IgG3-C_H1-IGF-1 and IgG3-H-IGF-1 can be recovered from the capillary fraction between 0 and 5 h after injection. Control unconjugated IGF-1 showed very little brain targeting at either 30 min (0.02% ID) or 4 h than (0.005%) after-injection, and control IgG3 showed no brain uptake at any time point.

Uptake of the IgG3-IGF-2 fusion proteins also depended on their structure. Twenty-four hours following injection, similar amounts of IgG3-C_H1-IGF-1 (0.06% ID), IgG3-H-IGF-1 (0.06% ID), and IgG3-C_H3-IGF-1 (0.08% ID) were found in the brain parenchyma. However, as was seen for the IGF-1 fusion proteins, the kinetics of uptake differed for the three IGF-2 fusions. IgG3-C_H1-IGF-2 was taken up most rapidly, reaching peak parenchymal levels 1 h after injection, and consistent with this rapid uptake, none of the protein was associated with the capillary fraction. In contrast, the IgG3-H-IGF-2 fusion protein was delayed in its uptake: 1 h after injection, it was found associated with the capillary pellet, but not in the brain parenchyma. However, by 4 h after injection, some uptake (0.04% ID) had occurred. IgG3-C_H3-IGF-2 was intermediate in its kinetics of uptake: at 1 h after injection, the protein (approximately 0.025% ID) was associated with both the capillary pellet and the brain parenchyma; however, by 4 h after injection, IgG3-C_H3-IGF-2 (approximately 0.07% ID) was detected only in the brain parenchyma.

The IgG3-Tf fusion proteins showed the most promise of all the fusions for use in brain targeting. Initial uptake kinetics (1 h after injection) were similar for IgG3-C_H3-Tf and IgG3-C_H1-Tf: both proteins showed approximately 0.10–0.15% ID in brain parenchyma and no uptake in the brain capillaries. However, by 24 h after injection, parenchymal levels of IgG3-C_H3-Tf remained at 0.15% ID, whereas levels of IgG3-C_H1-Tf had fallen to approximately 0.07% ID. IgG3-H-Tf was significantly and rapidly taken up: 0.3% of the ID was found in the brain 15 min after injection, and this level decreased only slightly (0.25% ID) by 24 h after injection. Thus, of all the fusion proteins, H-Tf achieved the highest concentration in brain parenchyma. Moreover, the kinetics were similar to, and the uptake was greater than that of control Tf, which showed rapid uptake of less than 0.1% of the ID.

The studies described above indicated that the radioactivity associated with the IgG3-Tf fusions exhibited significant uptake by the brain parenchyma. In order to confirm that this radioactivity represented intact fusion protein, recovery experiments were undertaken. For these experiments, ^{125}I-labeled IgG3-C_H3-Tf was infused through the right internal carotid artery and brain parenchymal fractions were prepared as described above, Radioactivity from either the perfusate or the parenchyma was immunoprecipitated with either DNS-Sepharose or protein G and analyzed by nonreducing SDS-PAGE. A 330-kD protein, representing the tetrameric form of IgG3-C_H3-Tf was observed in the brain parenchyma following both immunoprecipitation procedures, indicating that the proteins crossed the BBB as intact molecules.

In contrast to the anti-TfR antibody [15], the IgG3-Tf fusion proteins exhibit kinetics of uptake into brain parenchyma similar to Tf and do not accumulate in the brain capillaries. At the earliest times tested, the majority of the recovered protein was in the brain parenchyma, suggesting that these molecules rapidly transit across the BBB. Together the data reported above indicate the GF fusion proteins, especially those containing Tf, have potential for use in brain-targeting therapies.

IgG3-IL-2 Fusion Proteins

We now turn our attention to an IgG-cytokine fusion protein designed for use as an anti-tumor therapeutic agent. IL-2 is a 15-kD cytokine which stimulates T cells to proliferate and become cytotoxic [28]. IL-2 also stimulates natural killer cells and generates lymphokine-activated killer (LAK) cells, both of which respond with increased cytotoxicity toward tumor cells [29]. In addition, IL-2 stimulates resident, inactive tumor infiltrating lymphocytes to proliferate and become cytotoxic for the tumor [28–31]. These properties suggest that IL-2 could be effective in the treatment of cancer. However, these promising in vitro activities are difficult to apply in vivo to cancer patients because systemic IL-2 administration leads to increased vascular permeability resulting in vascular leak syndrome [32–34]. Additionally, because of the short in vivo half-life of IL-2 either multiple injections or continuous infusion are required to maintain an effective concentration [35].

The characteristics described above suggest that IL-2 could be effective in the immunotherapy of tumors if methods could be developed to increase the effective local concentration and limit the generalized toxicity of the cytokine by selectively delivering it to the site of the tumor [36]. Indeed, tumor cell lines transfected with the IL-2 gene secrete IL-2, become highly immunogenic, and provide protection against later challenge with the parent, non-

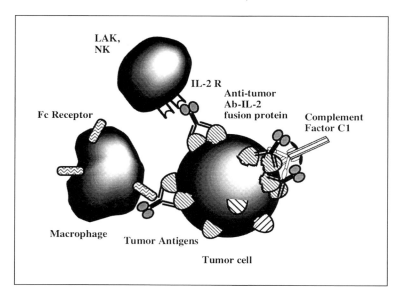

Fig. 4. Anti-tumor activities of Ab-IL-2 fusion proteins. The antibody variable region provides specific targeting to tumor antigens and increases the dose of IL-2 localized at the tumor. Ab-IL-2 fusions bound to both tumor-associated antigens and IL-2 Rs should cross-link effector cells with their tumor targets, improving cell-mediated killing. Simultaneous triggering of T cell function via IL-2 and Fc-mediated effector functions via the antibody could further improve immune stimulation.

immunogenic tumor [29]. These data demonstrate that sufficient concentrations of IL-2 at the site of a tumor can effectively stimulate the host immune system to identify and destroy the tumor cells; however transfection and reintroduction of tumor cells is difficult to apply in the clinical setting. Thus, the challenge is to develop an alternative approach for achieving effective local concentrations of IL-2.

Antibody-IL-2 fusion proteins possess a combination of properties that make them candidates for providing effective immune stimulation at the site of the tumor (fig. 4). First, the antibody variable region should provide specific targeting to tumor antigens, thereby increasing the dose of IL-2 localized at the tumor [37]. Second, IL-2 should activate the immune system to recognize a broad range of neoepitopes present on the transformed cells, thereby generating effective immunity in the face of antigenic modulation. Third, the short half-life of IL-2 should be extended by mAb conjugation due to both the increased size of the conjugate and the inherent stability of the mAb. Fourth, the anti-body-IL-2 fusion protein should enhance access to the tumor by increasing vascular permeability, as has been observed for both mAb-IL-2 chemical conjugates and IL-2/mAb combination therapy [38, 39]. Additionally, the simultaneous triggering of T cell function via IL-2 and Fc-mediated effector functions via the antibody could further improve immune stimulation. Finally, by simultaneously binding both tumor-associated antigens and IL-2 receptors (IL-2R), mAb-IL-2 fusions should cross-link effector cells with their tumor targets, thereby improving cell-mediated killing.

This reasoning has led various investigators to create IL-2 antibody fusion proteins. Although IL-2 can be used to replace the V_H of the antibody [6], our approach has been to create fusion molecules in which the nonimmunoglobulin partner is fused to IgG constant regions. Thus, a fusion protein linking human IL-2 to C_H3 of αDNS human IgG3 was engineered and expressed in myeloma cells [40]. Like IgG3, the IgG3-IL-2 fusion was assembled and secreted as an H_2L_2 molecule. The fusion protein migrated on SDS-PAGE gels with the expected molecular weight under both reducing and nonreducing conditions. The IgG3-IL-2 fusion was purified by affinity chromatography and tested for the ability to carry out both IL-2- and mAb-related activities.

The IgG3-IL-2 Fusion Protein Stimulates Both Proliferation and Cytotoxicity

When tested for the ability to stimulate proliferation of the IL-2-dependent murine cell line CTLL-2, half-maximal proliferation was achieved with IgG3-IL-2 concentrations (0.2–0.4 nM) that were only slightly higher than those required for human recombinant IL-2 (hrIL-2; 0.1–0.2 nM). These data demonstrated that the fusion protein retained the ability to stimulate T cell proliferation. Next, the ability of the fusion protein and hrIL-2 to generate LAK activity against ^{51}Cr-loaded target cells was determined. Results of these experiments showed that hrIL-2 generated maximal killing at 1,000 IU/ml and an effector to target (E/T) ratio of 50, whereas IgG3-IL-2 achieved maximal killing at 100 IU/ml and an E/T ratio of 50. In addition, IgG3-IL-2 consistently generated more efficient killing than a 10-fold higher concentration of hrIL-2 at all E/T ratios. Finally, at low concentrations (10 IU/ml) significant activity was generated for IgG3-IL-2, whereas very little activity was observed with IL-2.

The IgG3-IL-2 Fusion Protein Binds to High- and Intermediate-Affinity IL-2 Receptors

To further investigate the enhanced activities described above for the IgG3-IL-2 fusion protein, binding of hrIL-2 and IgG3-IL-2 to the IL-2R was examined [41]. The complex IL-2R is composed of three subunits with varying individual and combined affinities for IL-2 [reviewed in ref. 42]. The α-subunit (IL-2Rα) alone rapidly associates and dissociates with IL-2, and is referred to as the low-affinity receptor ($K_D = 1 \times 10^{-8}\ M$). The β- and γ-subunits comprise the intermediate-affinity receptor $K_D = 1 \times 10^{-9}\ M$), which is constitutively expressed on many cell types including T cells. Activation of T cells induces the expression of the α-subunit which combines with the β- and γ-subunits of the IAR to form the high-affinity receptor ($K_D = 1 \times 10^{-11}\ M$).

When cells expressing the intermediate-affinity receptor on their surface were used, half-maximal inhibition of the binding of ^{125}I-labeled IL-2 occurred at 3–4 nM for hrIL-2 and at 7–8 nM for IgG3-IL-2. In contrast, when cells expressing the high-affinity receptor were used, half-maximal inhibition occurred for hrIL-2 at 2–3 nM and for IgG3-IL-2 at 0.7–0.8 nM. These data suggested that IgG3-IL-2 has a significantly higher affinity than hrIL-2 for the high-affinity receptor. These two forms of the receptor differ by the presence or absence of the α-subunit. Therefore, we compared binding of the fusion protein and hrIL-2 to a soluble form of the α-subunit immobilized on the surface of an IAsys optical biosensor cuvette (Fisons Applied Sensor Technology). Analysis of the hrIL-2 binding data gave an association rate constant (k_a) of $7.0 \times 10^4\ M^{-1}\,s^{-1}$ and a dissociation rate constant (k_d) of $0.013\ s^{-1}$, which yield an equilibrium dissociation constant (K_D) of $1.9 \times 10^{-7}\ M$. Measurement of IgG3-IL-2 binding to sIL-2R α yielded a k_a of $7.5 \times 10^4\ M^{-1}\,s^{-1}$, which is similar to that for hrIL-2. Dissociation of IgG3-IL-2 differed significantly from that of hrIL-2. Greater than 60% of the bound IgG3-IL-2 remained bound even after repeated buffer washes over several hours. The fraction of the fusion protein that did dissociate exhibited a k_d of $0.004\ s^{-1}$. These k_a and k_d values were used to calculate a K_D of $5.3 \times 10^{-8}\ M$. These data indicate that the IgG3-IL-2 fusion protein exhibits at least a 20-fold higher binding affinity than native IL-2 for the α-subunit under these conditions.

IgG3-IL-2 Binds FcγRI and Activates C'

IgG3-IL-2 contains the Fc region of IgG3 and should possess many of the Fc-associated effector functions. Competition studies of ^{125}I-IgG3 binding to FcγRI showed that half-maximal inhibition of binding was obtained with 1–2 nM IgG3 or with 3–4 nM IgG3-IL-2, indicating that IgG3-IL-2 binds FcγRI with slightly decreased affinity compared to native IgG3. Although IgG3-IL-2 was able to direct C'-mediated lysis of antigen-coated SRBC in a dose-dependent manner, it was somewhat less effective than IgG3 assayed at similar concentrations.

In vivo Properties of IgG3-IL-2

The ultimate utility of an Ab-IL-2 fusion protein is dependent on the pharmacological properties of the molecule. Thus, in vivo properties of the IgG3-IL-2 fusion protein were compared to those of native IgG3 [Harvill and

Morrison, unpubl. data]. The biological half-life of ^{125}I-labeled IgG3-IL-2 injected intraperitoneally into BALB/c mice was approximately 8 h. This half-life is shorter than the 3 to 5-day half-life of IgG3, but is considerably longer-than the half-life reported for IL-2 [35]. Biodistribution of the injected IgG3 and IgG3-IL-2 also differed. Four hours after injection, the highest concentrations of IgG3 were found in the blood, whereas the highest concentrations of IgG3-IL-2 were found in the thymus, spleen, liver and lymph nodes, all of which are organs known to possess IL-2Rs. IL-2 has previously been shown to be cleared rapidly via the kidney [35]; however, in the current study, little IgG3-IL-2 was found in the kidney at 4 h. SDS-PAGE analysis of ^{125}I-labeled IgG3-IL-2 and ^{125}I-labeled IgG3 recovered from blood showed that both molecules remained intact 4 h following injection. Even though the fusion protein had an increased in vivo half-life, it did not cause serious toxicity. Intraperitoneal injection of larger quantities of fusion protein (100 µg) into mice did not result in any noticeable morbidity but did induce transient weight gain.

Effectiveness of the IgG3-IL-2 Fusion Protein in Eliciting a Tumor-Specific Immune Response

Previous studies in vivo with antitumor forms of an IgG1-IL-2 fusion protein showed that this protein can prevent the spread of human melanoma in human LAK-reconstituted SCID mice [43]. However, under these conditions, treatment with IgG1 alone showed a significant effect on tumor metastases, and treatment with either IgG1 plus IL-2 or the IgG1-IL-2 fusion protein had similar effects suggesting that there is no advantage of treatment with mAb-fusion proteins compared to combined treatment with the individual components. This model system differs from naturally occurring cancer in several major ways. A xenografted immune system may not be able to differentiate between cancerous and normal cells in the same way as a normal immune system. Additionally, LAK cell generation involves several days of in vitro treatment with high levels of IL-2 [31, 44].

To address these problems, we assessed the antitumor activity of the IgG3-IL-2 fusion protein using a syngeneic tumor in an animal with an intact immune system [Harvill and Morrison, unpubl. data]. For these experiments, we created an IgG3-IL-2 fusion protein containing variable regions specific for the idiotype of the antibody expressed on the surface of the B cell lymphoma 38C13 of the C3H/HeN mouse [45]. Anti-idiotypic (αId) antibodies specifically bind the surface immunoglobulin on the B cell lymphoma, providing a model system for evaluating the therapeutic potential of antibodies directed at tumor-associated antigens [46]. This tumor model has previously been used to demonstrate that antibodies of differing isotypes differ in their ability to affect tumor growth and that antibody and IL-2 administered together provide a more effective tumor treatment than does either administered alone [39, 47, 48].

In order to demonstrate the ability of the fusion protein to localize to the tumor in vivo, either ^{131}I-αId-IgG3-IL-2 or ^{131}I-αDNS-IgG3-IL-2 was injected into C3H/HeN mice bearing subcutaneous tumors on the right flank. Gamma camera imaging showed that the majority of labeled αId-IgG3-IL-2 present 24 h after injection was localized to the site of the tumor. In contrast, the small amount of the αDNS fusion protein remaining after 24 h was localized to the upper abdomen near the spleen and liver. Therefore, the tumor-specific portion of the fusion protein retains its ability to specifically localize to the site of the tumor. In these mice, the αId-IgG3-IL-2 exhibited a half-life of approximately 5 h, and approximately 12% of the injected radioactivity remained at the time of the gamma camera imaging.

To determine the effect of αId-IgG3-IL-2 on tumor growth in vivo, 1,000 38C13 cells were injected intraperitoneally into C3H/HeN mice. Groups of six mice were treated 24 h after tumor cell injection with either αId-IgG3 (10 μg), hr IL-2 (10,000 U), both αId-IgG3 (10 μg) and hrIL-2 (10,000 U), αId-IgG3-IL-2 (10 μg), or PBS as a control. Survival after treatment was followed over time. Three mice from the group treated with αId-IgG3-IL-2 and one treated with αId-IgG3 remained tumor-free throughout the study. All of the other mice died by day 35, resulting in a mean survival of approximately 25 days, irrespective of treatment.

The four surviving mice were subjected to a rechallenge experiment, as follows. Survivor mice as well as controls were treated by subcutaneous injection of 10,000 38C13 cells. All control mice died between 19 and 28 days following the rechallenge. All of the other rechallenged mice showed some resistance to tumor growth. The mouse previously treated with αId-IgG3 died 37 days after the rechallenge. One of the mice previously treated with αId-IgG3-IL-2 died on day 39. The second mouse previously treated with αId-IgG3-IL-2 remained tumor-free. The third mouse previously treated with αId-IgG3-IL-2 showed very slow tumor growth and survived for much longer. This mouse finally became moribund 60 days following the rechallenge and was sacrificed. Immunohistochemical analysis of the tumor from this mouse showed extensive infiltration of CD8+ cells. Some regions of the tumor also showed relatively high concentrations of CD4+ cells, and NK1.1+ cells were observed occasionally. Tumors from a naive mouse challenged with 10,000 38C13 cells contained no measurable CD8+, CD4+ or NK1.1+ cells. Therefore, in this mouse long-term survival and a decreased rate of tumor growth are associated with extensive infiltration of the tumor by immune cells, suggesting that a cytotoxic T lymphocyte response was in large part responsible for the extended survival. Consistent with this interpretation, nude mice subjected to the same treatment showed no increased survival.

A second experiment was conducted using a regimen that differed in two ways from that of the first experiment. First, the tumor cells were injected subcutaneously and treatment agents were injected peritoneally, allowing us to determine whether injection site altered the effectiveness of the IL-2 fusion protein. Second, lower treatment doses (2 µg) of fusion proteins and controls were administered. The same treatment groups were used as in the initial experiment, but each group contained ten mice. Tumors in mice treated with PBS or hrIL-2 grew rapidly and all of the mice in these two groups were dead by day 32. Mice treated with αId-IgG3 or αId-IgG3 and hrIL-2 showed extended survival (up to 55 days); however, all ultimately died. Mice treated with αId-IgG3-IL-2 showed extended survival and 2 of 10 mice remained tumor-free throughout the study.

The extended survival of fusion-protein treated mice in these studies suggests that an anti-tumor-IgG3-IL-2 fusion protein may be useful as a therapeutic agent against tumors. The beneficial effects of the fusion protein may be mediated by either the mAb or the IL-2 portion of the molecule. The importance of IL-2 stimulation of T cells in mediating tumor killing is suggested both by the nude mice experiment, in which no antitumor effects were observed, and by the rechallenge experiment, in which the tumor of the E2 mouse was shown to be extensively infiltrated by CD8+ T cells.

Potentiation of Host Antibody Response by αDNS-IgG3-IL-2-Bound Antigen

In addition to their antitumor activities, IL-2-Ag conjugates have been used in vaccines to increase the immune response to the attached antigens [49–53]. The mechanism for this effect is not well understood. IL-2 may target the attached antigen to particular cells for more effective antigen presentation, or may deliver a stimulatory signal that directly or indirectly enhances the immune response to the attached antigen. Physical linkage of the antigen to IL-2 has been shown to be critical for maximizing the immune response [49]. Because the hapten DNS can be easily linked to primary amine groups, the αDNS IgG3-IL-2 fusion protein described above may be useful for potentiating the immune response to any DNS-conjugated antigen.

In order to determine whether this hypothesis is correct, DNS-BSA-coated Sepharose beads were incubated with PBS, αDNS-IgG3 or αDNS-IgG3-IL-2 and injected into mice peritoneally. An enhanced antibody response to DNS was observed following treatment with αDNS-IgG3-IL-2, but not αDNS-IgG3. Increases in antigen-specific antibodies of all measured isotypes (IgA, IgM, IgG1, IgG2 and IgG3) were observed. The versatility of the DNS system provides a novel tool for the generation of IL-2-potentiated immune responses. This system should allow a panel of antigens to be tested rapidly to determine both the mechanism by which IL-2 affects the immune response, and to identify the protein, or peptides, that will be most useful in a therapeutic application.

The various studies described above demonstrate the utility of mAb-IL-2 fusion proteins. The IgG3-IL-2 fusion protein retains the activities of native IL-2 and IgG3, and in fact exhibits increased affinity for the IL-2 HAR. The significantly increased half-life of the fusion protein compared to HrIL-2 also provides a crucial therapeutic advantage. Treatment of tumor-bearing mice with an anti-tumor-IgG3-IL-2 fusion protein resulted in decreased tumor incidence and increased survival compared to controls. Finally, the αDNS-IgG3-IL-2 fusion protein may be useful for potentiating the immune response to a wide variety of antigens.

Polymeric IgG Molecules

We now turn our attention to a distinct group of novel engineered mAbs: IgGs constructed to form polymers using features normally associated with antibodies of the IgM isotype. IgM is distinguished by its ability to form disulfide-linked polymers: IgM pentamers consist of five IgM molecules covalently linked to the J chain [54], whereas IgM hexamers lack a J chain. The multiple Ag- and Fc-associated binding sites of polymeric IgM result in an increase in avidity for both antigen and effector molecules such as C'. However IgM, unlike IgG, neither binds to Fcγ [55, 56] nor exhibits a long serum half-life [57]. These observations led us to attempt to produce mAbs that combine the desirable qualities of IgM with those of IgG (fig. 5). Polymerization of IgM is thought to be dependent on both Cys414 in C_H3 in and an isotype-specific C-terminal peptide called the IgM tail piece (μtp) containing Cys575. Therefore, IgM-like IgGs were produced by engineering either or both of these features into the IgG3 heavy chain gene [58, 59].

Production of IgM-Like Mutants of IgG
We attempted to design polymeric IgGs by engineering Cys residues at sites in IgG corresponding to those in IgM. In initial experiments, Leu309 in C_H2 of IgG3, which corresponds by sequence homology to Cys414 in IgM (EU and OU numbering) [60] was mutated to Cys (IgG3L309C). Since there is no sequence homologous to the μtp in the IgG constant region, the entire 18-amino-acid μtp was joined to the end of the IgG3C_H3 (IgG3μtp). These two mutations were also combined in an IgG3 double mutant IgG3L309Cμtp. Additionally, μtp was transferred from IgG3 to IgGs 1, 2 and 4. For all of

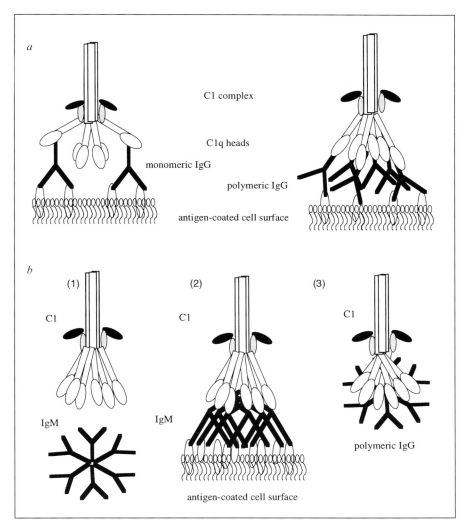

Fig. 5. Complement activation by polymeric IgM and IgGs. *a* At similar Ag densities, polymeric IgGs show increased ability to bind C1q and activate C′-mediated lysis compared to monomeric counterparts. *b* IgM does not activate C′ in the absence of Ag (1). Upon multivalent binding to antigen, IgM undergoes a conformational change rendering the C1q-binding sites accessible (2). In contrast, the aggregation of monomeric IgG resulting from Ag binding achieves C1q binding. Thus, polymeric IgGs with multiple associated Fcs are able to activate C′ in the absence of Ag (3).

these mutants, the variable regions were specific for DNS, and as described above, the mutant heavy chains were expressed with the DNS-specific chimeric light chain in non-producing murine myeloma cells, and proteins were purified by affinity chromatography.

All of the genetic mutations described above resulted in the synthesis of a mixed population of oligomeric antibodies ranging in size from monomers to hexamers. However, the mutants differ significantly in the proportions of oligomers present. The predominant form of freshly isolated IgG3L309C was monomer, but upon storage more of the higher-molecular-weight oligomers were seen. For IgG3L309Cµtp, the predominant forms were high-molecular-weight polymers and monomers. The IgG1, 2, 3, and 4 µtp forms were secreted primarily as polymers and monomers, but small amounts of intermediate forms were also present. There was no evidence for polymers containing more than six H_2L_2 subunits. This may be due to the natural tendency of IgG Fcs to form hexameric arrays through noncovalent interactions as observed on antigen-loaded monolayers [61]. An α-DNS IgM produced in the same cell line is secreted only as a fully assembled polymer, suggesting that some additional feature of IgM ensures complete polymerization. J chain, a protein of approximately 18 kD which has been implicated in the formation of pentamers, but not hexamers [54], was absent from all of the IgG-µtp polymers, but was present in IgM.

Analysis of Complement Activation by Polymeric IgG

Polymerization of IgG through Ag-specific aggregation is thought to be an important step in the activation of C'. Thus, the abilities of the IgG polymers and wild-type monomers to lyse Ag-coated SRBC were compared. All of the IgGµtp polymers exhibited increased C' activation ability compared to their monomeric counterparts and were virtually indistinguishable from IgM. In this assay, monomeric IgG3 consistently performs better than monomeric IgG1, whereas monomeric IgG2 shows little or no activity and monomeric IgG4 exhibits no activity. In fact, it is generally accepted that IgG4 is devoid of any C' activation ability [24, 62]. It was therefore surprising that IgG4µtp directs C'-mediated lysis of Ag-coated SRBCs almost as efficiently as the other IgGµtp isotypes. Fractionation of IgG3L309C into low-molecular-weight, mid-molecular-weight and high-molecular-weight forms showed a correlation between then level of C' activity and the extent of polymerization.

Next, the ability of the IgG-µtp proteins to bind C1q was assessed by an immunoassay. In this assay we typically observed that monomeric IgG3 exhibits greater C1q binding than IgG1, whereas IgG2 and IgG4 failed to bind C1q. All of the polymeric IgGs, however, exhibited increased C1q binding compared to their monomeric counterparts. Moreover, IgG1µtp, IgG2µtp and

activate C' and increased binding to the low-affinity FcRs present on immune effector cell. Indeed, we have noted differences in expression and effector function of our polymeric IgGs compared to both IgM and monomeric IgG. In IgM the µtp provides the penultimate cysteine required for covalent attachment of J chain. In myeloma expression system, IgM but not polymeric IgG contains J chain; therefore some additional feature of IgM must be required for J chain association. Also, as we had noted previously, secretion of IgG-µtp is not regulated in the same way as is secretion of IgM [58]: IgM is secreted only as pentamers and hexamers whereas the IgGµtp proteins are secreted as pentamers, hexamers, lower-molecular-weight polymers (dimers), and monomers. It has been proposed that binding of the immunoglobulin chaperonin BiP to the µtp retains IgM within the cell until polymerization is completed [76]. However, our results suggest that a µtp alone is not sufficient and that BiP, like J chain, requires some additional feature present in IgM but not in human IgG for association.

Perhaps the most surprising result we obtained with the IgGµtp polymers is that IgG4µtp is capable of directing C'-mediated lysis almost as effectively as IgM and the other IgGµtp isotypes. Although C' fixation by IgG4 Fc fragments aggregated on polystyrene latex beads had been reported [77], other investigators have found IgG4 bound to Ag incapable of C' activation [78, 79] and had partially mapped the residues responsible for this inactivity to C_H2 [23, 24]. Because monomeric IgG has only low affinity for C1q, C' activation by IgG requires formation of Fc arrays capable of simultaneously binding more than one of the six heads on the C1q molecule [80]. Formation of these arrays occurs spontaneously through noncovalent interactions of neighboring mAbs [61] and requires sufficient epitope density on the target surface. It is possible that monomeric IgG4 is incapable of these interactions and therefore fails to form Fc arrays, whereas IgG4µtp, forced to associate through the disulfide linkage of its µtp, presents a sufficient number of C1q binding sites in close enough proximity for C' activation to occur. Alternatively, the affinity of IgG4 for C1q may be significantly lower than that of the other IgGs; thus polymer formation may allow C' activation by altering the ability of IgG4 to bind C1q. Although IgG4µtp was effective in mediating lysis of Ag-coated target cells, this polymer was unable to consume C' when incubated with soluble antigen. The differences we see between the two types of assays may indicate that the polymer Fc-Ag interaction appropriate for C' activation may occur more readily on the surface of cells than in immune complexes.

The IgG polymers reached maximum C1q binding at much lower concentrations than wild type IgG, suggesting that increased avidity for the first component of C' is at least involved in, if not wholly responsible for, the increased C' activity of the engineered mAbs. However, IgM bound C1q in the range of wild type IgG and not at the lower antibody concentrations sufficient for polymeric IgG binding of C1q. Perhaps the advantage IgM has over IgG relates more to its ability to activate bound C1 rather than to its ability to first bind C1q. Alternatively, it is possible that the low level of immobilized antigen on the plate resulted in only mono- or bivalent binding that was sufficient for IgG to bind C1q, but insufficient to place IgM in a C1q binding conformation. Our findings support the concept that IgG Fc aggregation, no matter how it is achieved, is sufficient to activate the C′ cascade. Indeed, we found that the polymeric IgGs, with their multiple associated Fcs are able to activate C′ in the absence of Ag. The impact of this Ag-independent C′ activity on possible immunotherapeutic strategies remains to be determined.

The half-life of IgG polymers injected into mice was biphasic: the rapid α-phase appears to represent clearance of polymers, whereas the slower β-phase appears to represent clearance of monomers. Since IgGμtps do not contain a J chain, its presence is not responsible for the rapid clearance of the polymers. It is possible that the rapid clearance of some polymers is mediated through interactions with the low-affinity Fc receptors present on a number of lymphoid and phagocytic cells. It remains to be determined whether their rapid clearance will limit or enhance the in vivo usefulness of the polymeric IgGs.

The data reviewed herein readily demonstrate that numerous sophisticated techniques are now available to design and produce novel proteins. The challenge we now face is to harness the abundant information gained from these studies for rational design of antibodies with the desired combination of binding specificities and biological properties.

Acknowledgments

This work is supported by grants CA 16858, AI 29470, and AI39187 from the National Institutes of Health, and by grants IM-550 and IM-603 from the American Cancer Society. S.U.S. was supported in part by NINDS Training grant 1T32-NSO7356. R.I.F.S. was supported in part by DHHS Training Grants GM 07104 and Gm 08375. E.T. H. was supported in part by DHHS Training Grant T32 CA 09056. M.G.S. was supported in part by DHHS Training Grant CA 09120.

References

1   Ashkenazi A, Capon DJ, Ward RH: Immunoadhesins. Int Rev Immunol 1993;10:219–227.
2   Capon DJ, Chamow SM, Mordenti J, Marsters SA, Gregory T, Mitsuya H, Byrn RA, Lucas C, Wurm FM, Groopman JE, Broder S, Smith DH: Designing CD4 immunoadhesins for AIDS therapy. Nature 1989;337:525–531.

3   Traunecker A, Schneider J, Kiefer H, Karjalainen K: Highly efficient neutralization of HIV with recombinant CD4-immunoglobulin molecules. Nature 1989;339:68–70.

4   Byrn RA, Mordenti J, Lucas C, Smith D, Marsters SA, Johnson JS, Cossum P, Chamow SM, Wurm FM, Gregory T, Groopman JE, Capon DJ: Biological properties of a CD4 immunoadhesin. Nature 1990;344:667–670.

5   Martin S, Casasnovas JM, Staunton DE, Springer TA: Efficient neutralization and disruption of rhinovirus by chimeric ICAM-1/immunoglobulin molecules. J Virol 1993;67:3561–3568.

6   Landolfi NF: A chimeric IL-2/Ig molecule posseses the functional activity of both proteins. J Immunol 1991;146:915–919.

7   Schnee JM, Runge MS, Matsueda GR, Hudson NW, Seidman JG, Haber E, Quertermous T: Construction and expression of a recombinant antibody-targeted plasminogen activator. Proc Natl Acad Sci USA 1987;84:6904–6908.

8   Hoogenboom HR, Volckaert G, Raus JC: Construction and expression of antibody-tumor necrosis factor fusion proteins. Mol Immunol 1991;28:1027–1037.

9   Gillies SD, Young D, Lo KM, Foley SF, Reisfeld RA: Expression of genetically engineered immunoconjugates of lymphotoxin and a chimeric anti-ganglioside GD2 antibody. Hybridoma 1991;10:347–356.

10 Duffy KR, Pardridge WM: Blood-brain barrier transcytosis of insulin in developing rabbits. Brain Res 1987;420:32–38.

11 Fishman JB, Rubin JB, Handrahan JV, Connor JR, Fine RE: Receptor-mediated transcytosis of transferrin across the blood-brain barrier. J Neurosci 1987;18:299–304.

12 Rosenfeld RG, Pham H, Keller BT, Borchardt RT, Pardridge WM: Demonstration and structural comparison of receptors for insulin-like growth factor-I and -II (IGF-I and -II) in brain and blood-brain barrier. Biochem Biophys Res Commun 1987;149:159–166.

13 Duffy KR, Pardridge WM, Rosenfeld RG: Human blood-brain barrier insulin-like growth factor receptor. Metabolism 1988;37:136–140.

14 Bullard DE, Bourdon M, Bigner DD: Comparison of various methods for delivering radiolabeled monoclonal antibody to normal rat brain. J Neurosurg 1984;61:901–911.

15 Friden PM, Walus LR, Musso GF, Taylor MA, Malfroy B, Starzyk RM: Anti-transferrin receptor antibody and antibody-drug conjugates cross the blood-brain barrier. Proc Natl Acad Sci USA 1991;88:4771–4775.

16 Friden PM, Walus LR, Watson P, Doctrow SR, Kozarich JW, Backman C, Bergman H, Hoffer B, Bloom F, Granholm AC: Blood-brain barrier penetration and in vivo activity of an NGF conjugate. Science 1993;259:373–377.

17 Kordower JH, Charles V, Bayer R, Bartus RT, Putney S, Walus LR, Friden PM: Intravenous administration of a transferrin receptor antibody-nerve growth factor conjugate prevents the degeneration of cholinergic striatal neurons in a model of Huntington disease. Proc Natl Acad Sci USA 1994;91:9077–9080.

18 Shin SU, Morrison SL: Expression and characterization of an antibody specificity joined to insulin-like growth factor 1: Potential applications for cellular targeting. Proc Natl Acad Sci USA 1990;87:5322–5326.

19 Shin SU, Friden P, Moran M, Morrison SL: Functional properties of antibody insulin-like growth factor fusion proteins. J Biol Chem 1994;269:4979–4985.

20 Shin SU, Friden P, Moran M, Olson T, Kang YS, Pardridge WM, Morrison SL: Transferrin-antibody fusion proteins are effective in brain targeting. Proc Natl Acad Sci USA 1995;92:2820–2824.

21 Tan LK, Shopes RJ, Oi VT, Morrison SL: Influence of the hinge region on complement activation, C1q binding, and segmental flexibility in chimeric human immunoglobulins (published erratum appears in Proc Natl Acad Sci USA 1991;88:5066). Proc Natl Acad Sci USA 1990;87:162–166.

22 Duncan AR, Winter G: The binding site for C1q on IgG. Nature 1988;332:738–740.

23 Tao MH, Canfield SM, Morrison SL: The differential ability of human IgG1 and IgG4 to activate complement is determined by the COOH-terminal sequence of the C_H2 domain. J Exp Med 1991;173:1025–1028.

24 Tao MH, Smith RI, Morrison SL: Structural features of human immunoglobulin G that determine isotype-specific differences in complement activation. J Exp Med 1993;178:661–667.

25   Duncan AR, Woof JM, Partridge LJ, Buton DR, Winter G: Localization of the binding site for the human high-affinity Fc receptor on IgG. Nature 1988;332:563–564.

26   Canfield SM, Morrison SL: The binding affinity of human IgG for its high affinity Fc receptor is determined by multiple amino acids in the C_H2 domain and is modulated by the hinge region. J Exp Med 1991;173:1483–1491.

27   Chappel MS, Isenman DE, Everett M, Xu YY, Dorrington KJ, Klein MH: Identification of the Fc gamma receptor class I binding site in human IgG through the use of recombinant IgG1/IgG2 hybrid and point-mutated antibodies. Proc Natl Acad Sci USA 1991;88:9036–9040.

28   Lotze MT, Grimm EA, Mazumder A, Strausser JL, Rosenberg SA: Lysis of fresh and cultured autologous tumor by human lymphocytes cultured in T-cell growth factor. Cancer Res 1981;41: 4420–4425.

29   Grimm EA, Mazumder A, Zhang HZ, Rosenberg SA: Lymphokine-activated killer cell phenomenon. Lysis of natural killer-resistant fresh solid tumor cells by interleukin 2-activated autologous human peripheral blood lymphocytes. J Exp Med 1982;155:1823–1841.

30   Yron I, Wood TA Jr, Spiess PJ, Rosenberg SA: In vitro growth of murine T cells. V. The isolation and growth of lymphoid cells infiltrating syngeneic solid tumors. J Immunol 1980;125:238–245.

31   Hank JA, Surfus J, Gan J, Chew TL, Hong R, Tans K, Reisfeld R, Seeger RC, Reynolds CP, Bauer M, Wiersma S, Hammond D, Sondel P: Treatment of neuroblastoma patients with antiganglioside GD2 antibody plus interleukin-2 induces antibody-dependent cellular cytotoxicity against neuroblastoma detected in vitro. J Immunother 1994;15:29–37.

32   Ponce P, Cruz J, Travassos J, Moreira P, Oliveira J, Melo-Gomes E, Gouveia J: Renal toxicity mediated by continuous infusion of recombinant interleukin-2. Nephron 1993;64:114–118.

33   Siegel JP, Puri RK: Interleukin-2 toxicity. J Clin Oncol 1991;9:694–704.

34   Vial T, Descotes J: Clinical toxicity of interleukin-2. Drug Safety 1992;7:417–433.

35   Donohue JH, Rosenberg SA: The fate of interleukin-2 after in vivo administration. J Immunol 1983;130:2203–2208.

36   Rosenberg SA, Yang JC, Topalian SL, Schwartzentruber DJ, Weber JS, Parkinson DR, Seipp CA, Einhorn JH, White DE: Treatment of 283 consecutive patients with metastatic melanoma or renal cell cancer using high-dose bolus interleukin 2 (see comments). JAMA 1994;271:907–913.

37   Riethmuller G, Schneider GE, Johnson JP: Monoclonal antibodies in cancer therapy. Curr Opin Immunol 1993;5:732–739.

38   LeBerthon B, Khwali LA, Alauddin M, Miller GK, Charak BS, Mazumder A, Epstein AL: Enhanced tumor uptake of macromolecules induced by a novel vasoactive interleukin 2 immunoconjugate. Cancer Res 1991;51:2694–2698.

39   Berinstein N, Levy R: Treatment of a murine B cell lymphoma with monoclonal antibodies and IL2. J Immunol 1987;139:971–976.

40   Harvill ET, Morrison SL: An IgG3-IL2 fusion protein activates complement, binds FcγRI, generates LAK activity and shows enhanced binding to the high affinity IL-2R. Immunotechnology 1995;1: 95–105.

41   Harvill ET, Morrison SL: An IgG3-IL-2 fusion protein has higher affinity than hrIL-2 for the IL-2R alpha subunit: Real time measurement of ligand binding. Mol Immunol, in press.

42   Minami Y, Kono T, Miyazaki T, Taniguchi T: The IL-2 receptor complex: Its structure, function, and target genes. Annu Rev Immunol 1993;11:245–268.

43   Sabzevari H, Gillies SD, Mueller BM, Pancook JD, Reisfeld RA: A recombinant antibody-interleukin 2 fusion protein suppresses growth of hepatic human neuroblastoma metastases in severe combined immunodeficiency mice. Proc Natl Acad Sci USA 1994;91:9626–9630.

44   Hank JA, Robinson RR, Surfus J, Mueller BM, Reisfled RA, Cheung NK, Sondel PM: Augmentation of antibody dependent cell mediated cytotoxicity following in vivo therapy with recombinant interleukin 2. Cancer Res 1990;50:5234–5239.

45   Bergman Y, Haimovich J: Characterization of a carcinogen-induced murine B lymphocyte cell line of C3H/eB origin. Eur J Immunol 1977;7:413–417.

46   Maloney DG, Kaminski MS, Burowski D, Haimovich J, Levy R: Monoclonal anti-idiotype antibodies against the murine B cell lymphoma 38C13: Characterization and use as probes for the biology of the tumor in vivo and in vitro. Hybridoma 1985;4:191–209.

47   Kaminski MS, Kitamura K, Maloney DG, Campbell MJ, Levy R: Importance of antibody isotype in monoclonal anti-idiotype therapy of a murine B cell lymphoma. A study of hybridoma class switch variants. J Immunol 1986;136:1123–1130.

48   Berinstein N, Starnes CO, Levy R: Specific enhancement of the therapeutic effect of anti-idiotype antibodies on a murine B cell lymphoma by IL-2. J Immunol 1988;140:2839–2845.

49   Chen TT, Tao MH, Levy R: Idiotype-cytokine fusion proteins as cancer vaccines. Relative efficacy of IL-2, IL-4, and granulocyte-macrophage colony-stimulating factor. J Immunol 1994;153:4775–4787.

50   Hazama M, Mayumi-Aono A, Asakawa N, Kuroda S, Hinuma S, Fujisawa Y: Adjuvant-independent enhanced immune responses to recombinant herpes simplex virus type I glycoprotein D by fusion with biologically active interleukin-2. Vaccine 1993;11:629–636.

51   Hazama M, Mayumi AA, Miyazaki T, Hinuma S, Fujisawa Y: Intranasal immunization against herpes simplex virus infection by using a recombinant glycoprotein D fused with immunomodulating proteins, the B subunit of *Escherichia coli* heat-labile enterotoxin and interleukin-2. Immunology 1993;78:643–649.

52   Hinuma S, Hazama M, Mayumi A, Fujisawa Y: A novel strategy for converting recombinant viral protein into high immunogenic antigen. FEBS Lett 1991;288:138–142.

53   Nakao M, Hazama M, Mayumi AA, Hinuma S, Fujisawa Y: Immunotherapy of acute and recurrent herpes simplex virus type 2 infection with an adjuvant-free form of recombinant glycoprotein D-interleukin-2 fusion protein. J Infect Dis 1994;169:787–791.

54   Davis AC, Roux KH, Shulman MJ: On the structure of polymeric IgM. Eur J Immunol 1988;18:1001–1008.

55   Shenoy AM, Brahmi Z: Human IgG1 and mouse IgG3 but not monomeric IgM or IgE facilitate antibody-dependent cell-mediated cytotoxicity by human natural killer cells. Nat Immun Cell Growth Regul 1989;8:338–348.

56   Bruggemann M, Teale C, Clark M, Bindon C, Waldmann H: A matched set of rat/mouse chimeric antibodies. Identification and biological properties of rat H chain constant regions mu, gamma 1, gamma 2a, gamma 2b, gamma 2c, epsilon, and alpha. J Immunol 1989;142:3145–3150.

57   Vieira P, Rajewsky K: The half-lives of serum immunoglobulins in adult mice. Eur J Immunol 1988;18:313–316.

58   Smith RI, Morrison SL: Recombinant polymeric IgG: An approach to engineering more potent antibodies. Biotechnology 1994;12:683–688.

59   Smith RI, Coloma MJ, Morrison SL: Addition of a mu-tailpiece to IgG results in polymeric antibodies with enhanced effector functions including complement-mediated cytolysis by IgG4. J Immunol 1995;154:2226–2236.

60   Kabat EA, Wu TT, Perry HM, Gottesman KS, Foeller C: Sequence of proteins of immunological interest. NIH Publication No 91–3242, 1991.

61   Uzgiris EE, Kornberg RD: Two-dimensional crystallization technique for imaging macromolecules, with applicationn to antigen-antibody-complement complexes. Nature 1983;301:125–129.

62   Burton DR, Woof JM: Human antibody effector function. Adv Immunol 1992;51:1–84.

63   Anderson CL, Abraham GN: Characterization of the Fc receptor for IgG on a human macrophage cell line, U937. J Immunol 1980;125:2735–2741.

64   Lund J, Pound JD, Jones PT, Duncan AR, Bentley T, Goodall M, Levine BA, Jefferis R, Winter G: Multiple binding sites on the CH2 domain of IgG for mouse Fc gamma R11. Mol Immunol 1992;29:53–59.

65   Ravetch JV, Kinet JP: Fc receptors. Annu Rev Immunol 1991;9:457–492.

66   Feinstein A, Richareson NE, Gorick BD, Hughes-Jones NC: Immunoglobulin M conformational change is a signal for complement activation; in Celada, Shumaker and Sercarz (ed): Immunoglobulin M conformational change is a signal for complement activation. Plenum Press, NY, 1983, pp 47–57.

67   Goldstein GW, Betz AL: The blood-brain barrier. Sci Am 1986;255:74–83.

68   Frank HJ, Pardridge WM, Morris WL, Rosenfeld RG, Choi TB: Binding and internalization of insulin and insulin-like growth factors by isolated brain microvessels. Diabetes 1986;35:654–661.

69   Huebers HA, Finch CA: The physiology of transferrin and transferrin receptors. Physiol Rev 1987;67:520–582.

70   Pardridge WM, Eisenberg J, Yang J: Human blood-brain barrier transferrin receptor. Metabolism 1987;36:892–895.
71   Bayne ML, Applebaum J, Chicchi GG, Hayes NS, Green NG, Cascieri MA: Structural analogs of human insulin-like growth factor I with reduced affinity for serum binding proteins and the type 2 insulin-like growth factor receptor. J Biol Chem 1988;263:6233–6239.
72   Cascieri MA, Saperstein R, Hayes NS, Green BG, Chicchi GG, Applebaum J, Bayne ML: Serum half-life and biological activity of mutants of human insulin-like growth factor I which do not bind to serum binding proteins. Endocrinology 1988;123:373–381.
73   Ballard FJ, Read LC, Francis GL, Bagley CJ, Wallace JC: Binding properties and biological potencies of insulin-like growth factors in L6 myoblasts. Biochem J 1986;233:223–230.
74   Szabo L, Mottershead DG, Ballard FJ, Wallace JC: The bovine insulin-like growth factor (IGF) binding protein purified from conditioned medium requires the N-terminal tripeptide in IGF-1 for binding. Biochem Biophys Res Commun 1988;151:207–214.
75   Walton PE, Baxter RC, Burleigh BD, Etherton TD: Purification of the serum acid-stable insulin-like growth factor binding protein from the pig (Sus scrofa). Comp Biochem Physiol [b] 1989;92:561–567.
76   Sitia R, Neuberger M, Alberini C, Bet P, Fra A, Valetti C, Williams G, Milstein C: Developmental regulation of IgM secretion: The role of the carboxy-terminal cysteine. Cell 1990;60:781–790.
77   Isenman DE, Dorrington KJ, Painter RH: The structure and function of immunoglobulin domains. II. The importance of interchain disulfide bonds and the possible role of molecular flexibility in the interaction between immunoglobulin G and complement. J Immunol 1975;114:1726–1729.
78   Dangl JL, Wensel TG, Morrison SL, Stryer L, Herzenberg LA, Oi VT: Segmental gelxibility and complement fixation of genetically engineered chimeric human, rabbit and mouse antibodies. EMBO J 1988;7:1989–1994.
79   Bruggemann M, Williams GT, Bindon CI, Clark MR, Walker MR, Jefferis R, Waldmann H, Neuberger MS: Comparison of the effector functions of human immunoglobulins using a matched set of chimeric antibodies. J Exp Med 1987;166:1351–1361.
80   Hughes-Jones NC, Gardner B: Reaction between the isolated globular sub-units of the complement component C1q and IgG-complexes. Mol Immunol 1979;16:697–701.

Dr. Sherie L. Morrison, Department of Microbiology and Molecular Genetics,
405 Hilgard Avenue, University of Califormia, Los Angeles, CA 90095 (USA)

Capra JD (ed): Antibody Engineering.
Chem Immunol. Basel, Karger, 1997, vol 65, pp 159–178

....

Ligand Function of Antigenized Antibodies Expressing the RGD Motif

Rosario Billetta, Paola Lanza, Maurizio Zanetti

Department of Medicine and Cancer Center, University of California San Diego,
La Jolla, Calif., USA

Knowledge of the genetic organization and antibody structure has contributed enormously to the development of strategies that allow us to manipulate immunoglobulin genes and generate novel molecules that have different functions and properties. In this review, we will focus on the principle and engineering of antigenized antibodies as a method to express conformationally constrained peptide sequences of biological ligands such as the adhesion-related motif Arg-Gly-Asp (RGD). This amino acid sequence is found in several adhesive proteins which are implicated in cell adhesion and play a relevant role in metastatic processes serving as an anchorage point for tumor cells.

Antibodies are tetrameric molecules consisting of two identical heavy (H) chains joined to two identical light (L) chains by disulfide bonds and held together by inter-H-chain disulfide linkages. Both H and L chains have a variable (V) and a constant (C) region. Antibodies have two functional elements: the Fab fragment, which includes H and L chain V regions (V_H and V_L regions, respectively) and the carboxy-terminal portion of C_H1, and the Fc fragment, which consists of the remaining C_H domains of the H chain. Fab is responsible for idiotype and antigen specificity, whereas the Fc determines isotype and effector functions.

X-ray crystal analyses [1, 2] have clarified the three-dimensional structure of the antibody molecule and its topology of sequence conservation. Framework regions (FRs) are organized as β-strands interconnected with hypervariable (HV) loops. FRs are arranged in a β-sheet sandwich (the 'immunoglobulin fold') which is filled with hydrophobic side chains, and an invariant disulfide bond links the opposing sheets. The constraints on the side chains necessary to preserve folding are sufficiently stringent to explain the sequence conservation of these regions. In contrast, it is well established that HV regions can drift in sequence and conformation with little or no effect on the structure of the FRs [3]. As discussed in the following sections, both antigen binding and idiotypy localize to discrete areas of variability within the V domains: HV regions or complementarity-determining regions (CDRs). The conformation of the six CDR loops, three in the H and three in the L chain, is the principal determinant of the architecture of the antigen-binding site and also the major contributor to antibody antigenicity. For at least five of the six loops of the two chains, a discrete repertoire of conformations (canonical structures) exists on which alternative combinations and side-chain diversity generate a wide range of binding sites and, consequently, antigenic structures [4, 5]. A notable exception is the CDR3 of the H chain, which apparently does not use patterns of main conformation that exist in the current structural immunoglobulin data base. In general, antigenic determinants of immunoglobulin molecules may be seen as convex sites with large polar side chains hallmarked by the ability to provide contact bonds of the ionic, hydrogen, and van der Waals types for receptors, ligands and anti-idiotypic antibodies.

In common malignancies, including colon, lung, breast cancer and leukemias, once the immunosurveillance has been overcome, a relatively small portion of the cancer cells, due to a step-wise accumulation of mutations affecting the genetic stability of these cells, acquires the ability to grow in an uncontrolled fashion and start to metastasize throughout the body. Critical to successful growth invasion and metastasis is the process of neovascularization that accompanies these events in tumors [6, 7]. Both the events of tumor cell migration through the endothelium and the angiogenetic process critically depend on adhesion interactions mediated by cell surface receptors (integrins) (expressed by both tumor and endothelial cells) and the proteins of the extracellular matrix [8–10].

Integrins critically involved in angiogenesis, such as $\alpha v \beta 3$ [11], can be used as an experimental model to formally demonstrate the possibility to specifically target cellular structures to stop tumor growth and spreading throughout the body. Antibody loops are a new way to express oligopeptides in a conformationally constrained fashion using the hypervariable loops of an antibody molecule as the scaffold for peptide expression. The studies presented herein will summarize our experience and that of other laboratories on antibodies antigenized with the RGD peptide motif. It will be shown that these new engineered macromolecular ligands (antigenized antibodies) and various forms of their fragments can efficiently interact with tumor cells and constitute a new avenue for preventive or therapeutic intervention in cancer.

The Process of Antigenization

Antigenized antibodies, as a new concept and working tool, are predicated on the observations that natural antibodies, and their CDR loops, can share short stretches of amino acid sequences with antigens and mimic the function of the nominal ligand at the level of the corresponding receptor [12, 13]. Antigenization of antibody is a new immunological concept [14] and a working tool that allows one to investigate directly the molecular basis of antigenicity and immunogenicity of antibody molecules. Antigenization is also a new, powerful way to conformationally constrain oligopeptides and limit their flexibility within β-loops of V domains. These are exposed, solvent-accessible areas at the surface of the antibody molecule characterized by both convexity and relative rigidity [15]. β-Loops are, to some extent, independent of the physicochemical constraints that maintain the packing of V_H and V_L β-sheets in a conserved framework [4]. Experiments in which the CDRs of one antibody were transplanted into another antibody molecule clearly indicated that CDRs can be manipulated without affecting the overall packing of the V domains [16–18]. We considered using FR β-sheets as a scaffold for CDR loops re-engineered to encompass peptide epitopes of proteins other than immunoglobulins for antigenization of antibody [14]. Conceptually, our approach is different from the swapping of CDR loops between two antibody molecules [16–20]. Antigenization is a new method to express a potentially vast repertoire of discrete peptide epitopes whose activity may require conformational constraining.

Materials and Methods

The method developed in our laboratory for antibody antigenization over the years consist in cassette mutagenesis [21], although other methods of antigenization can be utilized and have been reported [22–24]. The experiments reported herein were carried out in the CDR3 of the H chain, as in natural antibodies the CDR3 loop is the most variable in length and amino acid composition.

Computer-Based Analysis and Selection of the Engineered Sequence

Hydrophilic sites of proteins tend to be surface exposed, and antigenicity depends on immunological accessibility. Hydrophilicity analyses were generated using the Kyte-Doolittle algorithm [25] in the program Mac Vector™ 3.5. The window size used for the analysis was 3. Values above axis denote hydrophilic regions that are exposed on the outside of the molecule. Values below the axis indicate hydrophobic regions that tend to be buried inside the molecule or inside other hydrophobic regions. The Kyte-Doolittle algorithm was originally used for hydrophobicity profiles; in the Mac Vector 3.5 program the sign of the values has been reversed and the hydrophilicity is plotted instead. Sometimes, antigenic indexes of antigenized V regions are also used and derived with a combination of different algorithms:

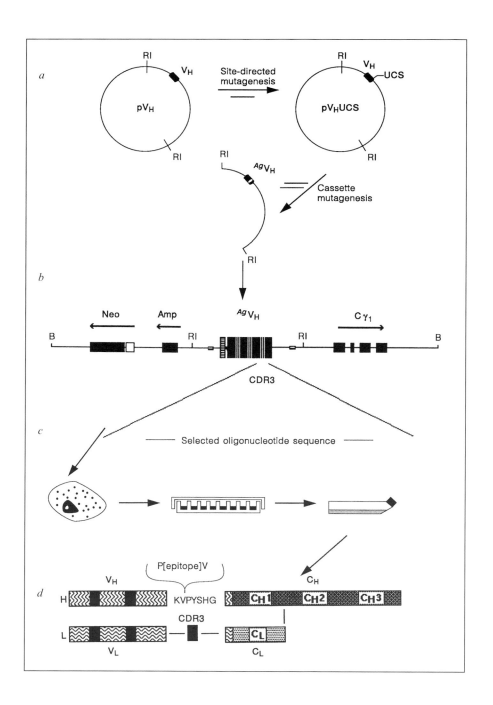

e.g., Kyte and Doolittle [25], Hopp and Woods [26], Goldman et al. [27] for hydrophilicity values, Janin et al. [28] and Emini et al. [29] for surface probability, Karplus and Schulz [30] for segmental flexibility along with the secondary structure predictions of Chou-Fasman and Robson-Garnier. These combined analyses produce a prediction of the surface contour of the protein and localize possible surface-exposed regions.

Computer-generated models of RGD and RGD$_3$ were built by using the program Homology (Biosym Technologies, San Diego, Calif., USA). Details on the modeling techniques can be found in Zanetti et al. [31].

Antigenization of Antibodies with the RGD Motif

V regions of antibodies are engineered according to the general strategy illustrated in figure 1. The engineering procedures are essentially the same for any oligopeptide sequence and possibly apply to any V region gene. The process of antigenization typically begins by creating a unique cloning site by site-directed mutagenesis into the V region loop(s), e.g. the CDR3 of the H chain, and allows the insertion of the sequence of interest within a selected location of the host V region gene. Deletion of the existing CDR is not necessary; however, this may be required when the sequence to be introduced is long (> 20 residues) and computational analysis suggests a poor surface exposure or a tendency to bend inward. The new sequence is introduced in the antibody loop to be antigenized by cloning a pair of specific complementary synthetic oligonucleotides into the unique cloning site of the mutagenesis vector. The DNA fragment carrying the engineered V_H gene is then subcloned upstream from a constant region expression vector. Alternative technical procedures have been used in other laboratories to obtain expression of a certain peptide in antibody loops. One is by PCR-mutagenesis [22],

Fig. 1. Schematic representation of the technical steps involved in the process of antigenization of antibody. *a* The plasmid pV$_H$ contains the *Eco*RI fragment carrying the genomic V$_H$ gene to be used for antibody antigenization. A unique cloning site (UCS) is created by site-directed mutagenesis at the level of one or multiple CDRs (in the example shown, the unique cloning site is created in the CDR3 of the V$_H$ gene). This originates an intermediate plasmid, pV$_H$UCS. A pair of complementary oligonucleotides coding for the peptide sequence of choice, and containing ends compatible with the unique cloning site, are annealed and then subcloned into the host V$_H$ gene. *b* The engineered *Eco*RI DNA fragment containing the antigenized V$_H$ gene (^{Ag}V$_H$) is then subcloned in the proper transcriptional orientation into an expression vector upstream from an immunoglobulin constant (C) region gene. B = *Bam*HI; RI = *Eco*RI; Neo = neomycin (G418) resistance; Amp = ampicillin resistance. *c* The final DNA construct is electroporated in a host cell line ($\sim 2 \times 10^7$ cells). Transfected cells are incubated without selection for 24 h and then selected in the presence of neomycin. Neomycin-resistant clones that secrete the antigenized antibody (^{Ag}Ab) are identified by ELISA and expanded by large-scale culture. ^{Ag}Abs are purified by affinity chromatography on a protein A column and eluted at acid pH. *d* Schematic representation of a prototypic ^{Ag}Ab. The H chain originates from the fusion of the ^{Ag}V$_H$ containing the selected epitope with an immunoglobulin C region. The VP doublets constitute the unique cloning site which flanks the inserted oligopeptide sequence in the following configuration: VP(epitope)VP. The L chain is either provided by the myeloma cells (single transfection procedure) or transfected (double-transfection procedure). H and L chains are not to scale. From Zanetti et al. [31], with permission.

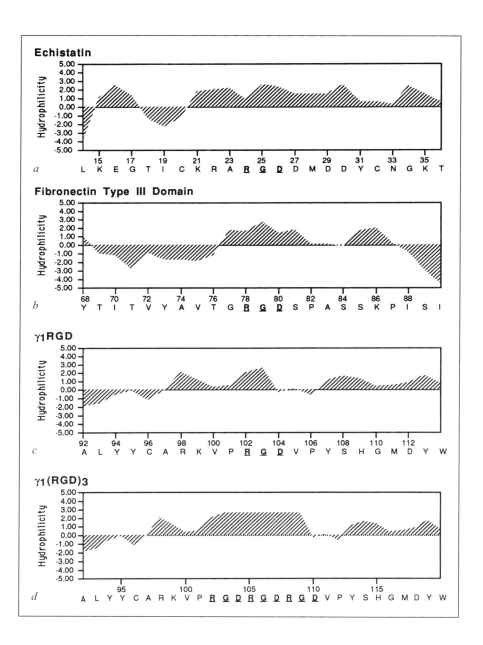

Echistatin

a   L K E G T I C K R A **R** **G** **D** D M D D Y C N G K T

Fibronectin Type III Domain

b   Y T I T V Y A V T G **R** **G** **D** S P A S S K P I S I

γ1RGD

c   A L Y Y C A R K V P **R** **G** **D** V P Y S H G M D Y W

γ1(RGD)₃

d   A L Y Y C A R K V P **R** **G** **D** **R** **G** **D** **R** **G** **D** V P Y S H G M D Y W

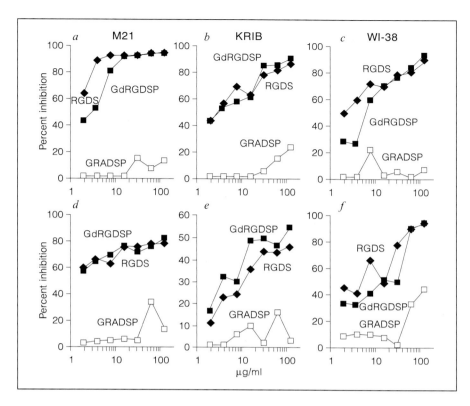

Fig. 5. Inhibition of adhesion to γ1(RGD)₃ to fibronectin using RGD-containing peptides. The cell lines were M21, KRIB and WI-38. Inhibition of adhesion was studied by incubating the cells in the substrate-coated plates γ1(RGD)₃ (*a–c*) or fibronectin (FN) (*d–f*) in the presence of various concentrations of inhibitor synthetic peptides for the duration (1 h at 37 °C) of the adhesion assay. Results were expressed as percent inhibition calculated as: (OD of cells in medium only – OD of cells in the presence of inhibitor/OD of cells in medium only) × 100. From Lanza et al. [39], with permission.

another study, Lee et al. [24] inserted the RGD sequence into a loop of the immunoglobulin V_L domain REI in a way to orient the RGD moiety into an active conformation. Unfolded or proteolytically fragmented forms of the antigenized REI L chain but not the intact molecule competed for binding to αIIbβ3, suggesting that the RGD sequences must be solvent accessible in order to react with the integrin receptor. The authors also engineered two REI proteins in which the CDR3 contained RGD flanked by residues found in the snake venom platelet antagonist kistrin. These antigenized REI variants competed strongly for fibrinogen binding.

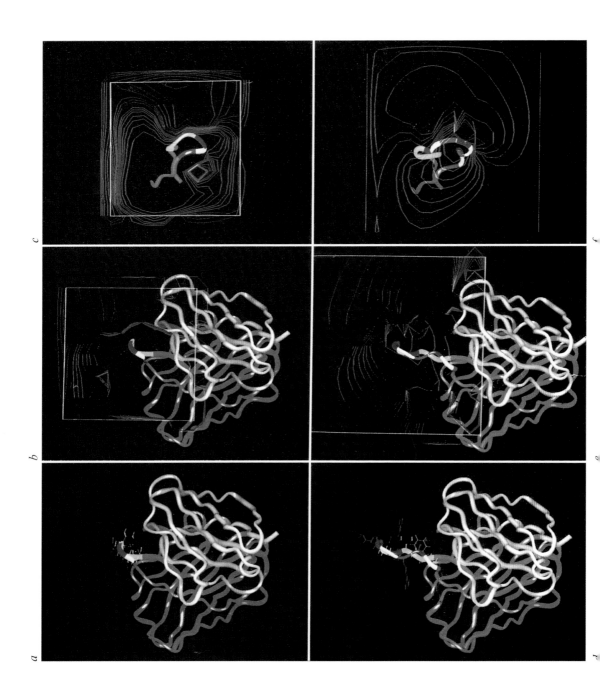

Another useful model of RGD-containing ligands is offered by comparative studies of murine monoclonal antibodies that bind the RGD recognition site of αIIbβ3. The specificity of each of the monoclonal antibodies PAC-1 [40, 41], OP-G2 [42, 43] and LJ-CP3 [44] resides essentially in the CDR3 containing the RYD sequence which is contributed by the immunoglobulin germ line D gene segment DSP 2.10 [45]. All these antibodies can compete with each other in binding to αIIbβ3, and this binding is inhibited by RGD-containing peptides.

Collectively, these studies indicate that antigenization with the RGD motif successfully recreated some of the ligand functions of adhesive proteins when expressed in the β-loops of antibodies. Additionally, these studies open the way to new strategies of immunointervention based on disruption of the RGD and other adhesive motif tumor cell/receptor(s) interaction, and the development of specific antagonists for example of the angiogenesis process, which may play a relevant role in future approaches for tumor therapy. With respect to our own studies, several considerations can be made. First is the fact that only the antibody expressing RGD repeated three times mediated adhesion of tumor cells and bound by flow cytometry. The antibody engineered with one RGD only and an antibody engineered to express a 12mer hydrophilic motif of different amino acid composition both failed to mediate tumor cell adhesion. Their binding to cells by flow cytometry was minimal if any. In each instance, adhesion was inhibited in a dose-dependent manner by synthetic peptides containing RGD, indicating that the γ1(RGD)$_3$-mediated effect(s) are RGD specific. Second, the difference in adhesion and cell-binding activity between γ1(RGD)$_3$ and γ1RGD implies that only the (RGD)$_3$ loop has the

Fig. 6. Molecular models of the engineered V regions RGD and (RGD)$_3$. The RGD-containing molecule is shown in the upper panels and (RGD)$_3$ in the lower panels. The L chain is pink, and the H chain is green. The antigenized loops are in yellow and single amino acids are denoted by different colors: arginine, blue; glycine, white; aspartic acid, red. *a, d* Molecules as solid ribbons through the backbone atoms with the (RGD)$_3$ loop projecting much farther from the main body of the V_H region than RGD. Both loops are fairly rigid, as shown by high-temperature molecular dynamics, and are roughly planar. The side chains of the loop residues point away from each other, accentuating the dipolar character of the loops. *b–e, c–f* Contours of the electrostatic fields of the two molecules. *b, e* Side views of the RGD and (RGD)$_3$ loops, respectively. *c, f* Top views. Positive contours are shown in blue, and negative contours are shown in red. Contour values are ±0.01, ±0.02, ±0.03, ±0.05, ±0.07, ±0.1, and ±0.15 kT/e. In the side view, RGD is largely negative, and (RGD)$_3$ has alternating positive and negative lobes. However, (RGD)$_3$ is also more positive at the N-terminal end of the loop and negative at the C-terminal end. In the top views, RGD is still largely negative, presumably because of the proximity effects of the rest of the protein. On the other hand, (RGD)$_3$ is markedly dipolar in this view. From Zanetti et al. [31], with permission.

appropriate conformation to interact with a receptor on tumor cells. Therefore, it appears that the mere fact of constraining a functional motif within an antibody loop is per se insufficient to confer biological activity to the antibody. Combination of computer-modeling-guided design of antibody [31] and selection of semicombinatorial display libraries randomized by PCR-directed mutagenesis [23] can be the approach for the identification of functional motifs with optimal biological activity.

Discussion

Considerations for Peptide Structure

Modern chemistry and immunology are aiming towards the definition of active sites of antigens, ligands and receptors in order to create simple and efficient ways to manipulate the immune response for disease prevention and/ or therapy. The process of drug design has been found to be arduous as well as technical, and logistical problems have been only partially solved. Antibody antigenization is a new powerful approach to express epitopes for their conformational analysis. By virtue of expression within the physicochemical constraints of the three-dimensional fold of an immunoglobulin V domain, oligopeptides acquire conformational stability. This implies that peptides will assume only a few of the many possible conformations allowed in solution. However, because much information already exists relative to antibodies [46], antigenized antibodies are also a formidable tool to study the conformation of selected portions of antigens for which information does not exist or for which conventional methods of structural analysis are not readily applicable. The models shown in figure 6 were built using the Homology program developed by Biosym Technologies. Modeling of the conformation of the RGD-containing loop was accomplished through a series of steps as described in detail in Zanetti et al. [31] and summarized herein in Material and Methods. The results of this analysis show that both RGD-containing loops project outward from the main body of the protein, ensuring that they are fully accessible for binding to a putative acceptor molecule. A comparison of figure 6a and 6d shows that the $(RGD)_3$-containing loop extends much farther into solvent. All low-energy conformations of the models of the $(RGD)_3$ loop appear to be quite similar, indicating that each loop has a fairly rigid structure, an effect probably due to residue Arg102, which is highly strained in the model, being crowded by the main body of the protein, mostly the H chain. The backbone structure of the loop also constrains the charged residues to be quite far apart (data not shown), so that oppositely charged side chains do not intertwine to neutralize any long-range electrostatic effect. Further analysis of the molecular model indicated that oligomerization of RGD in the antibody loop enhanced surface accessibility but also created interesting electrostatic profiles. Side views of the RGD and (RGD)₃ show their highly polar character. RGD is mostly negative, and (RGD)₃ has an alternating positive and negative pattern with a prevalently positive N-terminal and a prevalently negative C-terminal. This difference is most apparent when the loops are viewed from the top. This likely is the electrostatic profile presented to a putative surface acceptor molecule approaching from a distance. Interestingly, even in the shorter loop, the side chains of Arg102 and Asp104 point away from each other, implying the existence of a dipole. Although this was not reflected in the electrostatic field, the dipole may be lost in the field for the rest of the protein, and the effect can be seen only when the loop is large enough to project away. Thus, in the unbound state, the (RGD)₃ loop contains both positively and negatively charged residues in proximity. This study demonstrates that the highly charged, hydrophilic RGD motif is fully accessible at the surface of the molecule. The covalent attachments of the RGD peptide at the end points of two β-strands was effective in imposing a restrained conformation to the loop. Thus placing the antigen within the antibody loop has not decreased the presentation potential appreciably while stabilizing the conformation somewhat. The high density of charged amino acid residues in the loops gives these loops interesting electrostatic profiles with side views of the RGD and (RGD)₃ loops revealing their highly polar character.

In conclusion, it is worth mentioning that only the RGD₃-containing loop is the one that has appreciable biological activity in inhibiting NK activity [31] and promoting tumor cell adhesion [39]. This confirms the value of computer modeling for the design of molecules with significant biological activity and for peptide motif modeling.

Considerations for Tumor and Angiogenesis

Under normal circumstances immune defenses are not sufficient to stop tumor growth against transformed cell growth and spreading. Several factors account for this: immunosuppression, failure of tumor cells to express efficiently major histocompatibility antigens and costimulatory molecules like B7-1 and 2, immunoselection of tumor variants with reduced expression of tumor antigens, and the poor immunogenicity of tumor antigens often carbohydrates in nature.

The target of tumor immunotherapy is to either diminish or stop tumor growth at the primary site or limit its spreading to other organs. The first goal has been pursued using antibodies against tumor-associated antigens and more recently by activating T lymphocytes specific for tumor specific antigens (tumor rejection antigens) or virus antigens when viruses are the cause of malignant transformation. A rational approach to the second goal is to compete for molecularly defined structures used by tumor cells as attachment sites to normal tissues. In both instances, it will be required that measures of immunointervention be precise in molecular terms to ensure the maximum degree of specificity and efficacy.

Metastasis is an important event in tumorigenesis and makes cancers hard to eradicate surgically, pharmacologically or by local irradiation. Three main steps are necessary for tumor cells to cross the basal lamina in order to leave the tissue of origin and enter the circulation to eventually colonize other sites in the body. First of all, tumor cells express laminin receptors which allow the cells to adhere to the lamina, second they must secrete type IV collagenase to digest the basal lamina. Finally, there is migration through the endothelium which involves the interaction with several extracellular matrix proteins. Therefore integrin-mediated cell adhesion plays a role in many aspects of tumor invasion. Following this initial phase, the formation of independent tumor foci is favored by the release of tumor angiogenesis factors with vascularization of the ectopic tumor.

Angiogenesis is a multistep process, in which capillary endothelial cells sever their normal cell-cell attachments, migrate through the extracellular matrix and reform cell-cell attachments to create new capillaries. As the interactions of endothelial cells with each other and the underlying substratum are crucial to this process, the endothelial cell adhesion molecules that mediate cell-cell and cell-substratum are of fundamental importance to angiogenesis [8–10]. Angiogenesis is essential for tumor growth and progression.

Integrin $\alpha v \beta 3$, the RGD-dependent endothelial cell receptor for vonWillebrand factor, fibrinogen, vitronectin and fibronectin [8], is preferentially expressed on blood vessels undergoing angiogenesis and has been shown recently to play an important role in neovascularization [11]. The integrin $\alpha v \beta 3$ is a receptor that allows vascular endothelial cells to interact with a wide variety of extracellular matrix components. Antagonists of $\alpha v \beta 3$ have been showna to block the formation of new blood vessels without having any effect on preexisting vessels [11]. These antagonists have been shown to cause regression of human tumors of distinct histological origin, suggesting that antagonists of the $\alpha v \beta 3$ may be of therapeutic use against tumors.

As shown above, the hydrophilic amino acid sequence Arg-Gly-Asp (RGD) [47] mediates the adhesion of normal and tumor cells to components of the extracellular matrix. This amino acid sequence is found in several adhesive proteins including fibronectin, vitronectin, fibrinogen, von Willebrand factor and collagens [48–50]. In these molecules, the core RGD sequence is variably flanked by different residues and the total number of RGD triplets also varies.

Synthetic peptides containing RGD have been shown to inhibit tumor cell adhesion to extracellular matrix components in vitro [51]. In vivo when administered concomitantly with the tumor (murine melanoma cells) they have been shown to diminish metastasis formation in mice [52, 53]. This suggests that the receptor engaged in the interaction with RGD can be inhibited in vivo and plays a role in the dissemination of tumor cells in the blood stream.

Antigenized antibodies and their fragments offer several advantages over synthetic peptides or peptido-mimetics including a more stable expression of the peptide motif in a constrained fashion, slower clearance, and relative resistance to proteolysis in circulation. In summary antigenized antibodies and their fragments could be used to reach two main objectives: (1) By intercepting tumor cells and blocking the formation of additional metastatic foci, these molecules could serve as adjuvant to standard chemotherapy, a preventive measure against the spread of metastatic cells after surgical removal of primary tumor, and immunotherapeutic measure during the early phases of clinically documented metastatic processes. (2) By acting as integrin antagonists, they could specifically block the process of neovascularization, which has been shown to be crucial for tumor growth, invasion and metastasis, and therefore cause cancer regression. In conclusion, as reviewed above, molecular modeling of the antigenized antibody loops proved to be a method to predict the conformation of the newly built functional loops and model peptide structure. Structural prediction and refinement through the use of molecular design combined with selection methods can aid the engineering of new molecules with significant biological activity. The relationship between molecular modeling, immunological research and drug design will become more intertwined. Feedback from biological assays can then give information as to how accurate predictions of activity are. The use of antibody-based immunotherapeutics of novel formulation, such as antigenized antibodies and various forms of their fragments, expressing oligopeptides in a conformationally constrained fashion, can be foreseen as a method to develop novel molecules. Highly specific recombinant molecules, antagonist of specific cell surface receptors, have the potential to be used as adjuvant to standard chemotherapy as a measure to the spread of tumor cells and in other diseases.

Acknowledgments

This work was supported by the Council for Tobacco Research and NIH grant AI33204. M.Z. is Faculty Member of the Biomedical Sciences Graduate Program.

References

1 Schiffer M, Girling R, Ely K, Edmundson A: Structure of a λ-type Bence-Jones protein at 3.4 Å resolution. 1973;12:4620–4631.
2 Poljak RJ, Amzel LM, Avey HP, Chen BL, Phizackerly RP, Saul F: Three-dimensional structure of the Fab′ fragment of a human immunoglobulin at 2.8 Å resolution. Proc Natl Acad Sci USA 1973;70:3305–3310.
3 Kabat EA, Wu ET, Reidmiller M, Perry HM, Gottesman KS: Proteins of immunological interest. Bethesda, US Department of Health and Human Services, 1987.
4 Chothia C, Novotny J, Bruccoleri R, Karplus M: Domain association in immunoglobulin molecules: The packing of variable domains. J Mol Biol 1985;186:651–663.
5 Chothia C, Lesk AM: Canonical structures for the hypervariable regions of immunoglobulins. J Mol Biol 1987;196:901–917.
6 Dvorak HF: Tumors: Wounds that do not heal. Similarities between tumor stroma generation and wound healing. N Engl J Med 1986;315:1650–1659.
7 Weidner N, Folkman J, Pozza F, Bevilacqua P, Allred EN, Moore DH, Meli S, Gasparini G: Tumor angiogenesis: A new significant and independent prognostic indicator in early-stage breast carcinoma (see comments). J Natl Cancer Inst 1992;84:1875–1887.
8 Cheresh DA: Human endothelial cells synthesize and express an Arg-Gly-Asp-directed adhesion receptor involved in attachment to fibrinogen and von Willebrand factor. Proc Natl Acad Sci USA 1987;84:6471–6475.
9 Janat MF, Argraves WS, Liau G: Regulation of vascular smooth muscle cell integrin expression by transforming growth factor betal and by platelet-derived growth factor-BB. J Cell Physiol 1992;151:588–595.
10 Cheng YF, Kramer RH: Human microvascular endothelial cells express integrin-related complexes that mediate adhesion to the extracellular matrix. J Cell Physiol 1989;139:275–286.
11 Brooks PC, Clark RA, Cheresh DA: Requirement of vascular integrin alpha v beta 3 for angiogenesis. Science 1994;264:569–571.
12 Ollier PR-S, Somme JG, Theze J, Fougereau M: The idiotypic network and the internal image: Possible regulation of a germline network of paucigene encoded Ab2 (anti-idiotypic) antibodies in the GAT system. EMBO J 1985;4:3681–3688.
13 Bruck C, Co MS, Slaoui M, Gaulton GM, Smith T, Fields BM, Maulon GI, Greene MI: Nucleic acid sequence of an internal image-bearing monoclonal anti-idiotype and its comparison to the sequence of the external antigen. Proc Natl Acad Sci USA 1986;83:6578–6582.
14 Zanetti M, Billetta R, Sollazzo M: Idiotype vaccines by antibody engineering. Structural and functional considerations; in Cazenave PA (ed): Anti-Idiotypic Vaccines. New York, Springer-Verlag, 1990.
15 Novotny J, Tonegawa S, Saito H, Kranz DM, Eisen HN: Secondary, tertiary, and quaternary structure of T-cell-specific immunoglobulin-like polypeptide chains. Immunol Today 1986;83:742–746.
16 Verhoeyen M, Milstein C, Winter G: Reshaping human antibodies: Grafting an antilysozyme activity. Science 1988;239:1534–1536.
17 Jones P, Dear P, Foote J, Neuberger M, Winter G: Replacing the complementary-determining regions in a human antibody with those from a mouse. Nature 1986;321:522–524.
18 Riechmann L, Clark M, Waldmann H, Winter G: Reshaping human antibodies for therapy. Nature 1988;332:323–327.
19 Tempest PR, Bremner P, Lambert M, Taylor G, Furze JM, Carr FJ, Harris WJ: Reshaping a human monoclonal antibody to inhibit human respiratory syncytial virus infection in vivo. Bio/Technology 1991;9:266–271.
20 Co MS, Deschamps M, Whitley RJ, Queen C: Humanized antibodies for antiviral therapy. Proc Natl Acad Sci USA 1991;88:2869–2873.
21 Sollazzo M, Billetta R, Zanetti M: Expression of an exogenous peptide epitope genetically engineered in the variable domain of an immunoglobulin: Implications for antibody and peptide folding. Protein Eng 1990;4:215–220.

22   Zaghouani H, Krystal M, Kuzu H, Moran T, Shah H, Kuzu Y, Schulman J, Bona C: Cells expressing an H chain Ig gene carrying a viral T cell epitope are lysed by specific cytolytic T cells. 1992;148:3604–3609.

23   Barbas CD, Languino LR, Smith JW: High-affinity self-reactive human antibodies by design and selection: Targeting the integrin ligand binding site. Proc Natl Acad Sci USA 1993;90:10003–10007.

24   Lee G, Chan W, Hurle MR, DesJarlais RL, Watson F, Sathe GM, Wetzel R: Strong inhibition of fibrinogen binding to platelet receptor alpha IIb beta 3 by RGD sequences installed into a presentation scaffold. Protein Eng 1993;6:745–754.

25   Kyte J, Doolittle R: A simple method for displaying the hydrophilic character of a protein. J Mol Biol 1982;157:105–132.

26   Hopp T, Woods K: Prediction of protein antigenic determinants from amino acid sequences. Proc Natl Acad Sci 1981;78:3824–3828.

27   Engelman D, Steitz T, Goldman A: Identifying nonpolar transbilayer helices in amino acid sequences of membrane proteins. Annu Rev Biophys Biophys Chem 1986;15:321–353.

28   Janin J, Wodak S, Levitt M, Maigret B: Conformation of amino acid side-chains in proteins. J Mol Biol 1978;125:357–386.

29   Emini E, Hughes J, Perlow D, Boger J: Induction of hepatitis A virus-neutralizing antibody by a virus-specific synthetic peptide. J Virol 1985;55:836–839.

30   Karplus P, Schulz G: Prediction of chain flexibility in proteins. A tool for the selection of peptide antigens. Naturwissenschaften 1985;72:212–213.

31   Zanetti M, Filaci G, Lee RH, del Guercio P, Rossi F, Bacchetta R, Stevenson F, Barnaba V, Billetta R: Expression of conformationally constrained adhesion peptide in an antibody CDR loop and inhibition of natural killer cell cytotoxic activity by an antibody antigenized with the RGD motif. EMBO J 1993;12:4375–4384.

32   Morrison S: Transfectomas provide novel chimeric antibodies. Science 1985;229:1202–1207.

33   Main A, Harvey T, Baron M, Boyd J, Campbell I: The three-dimensional structure of the tenth type III module of fibronectin: An insight into RGD-mediated interactions. Cell 1992;71:671–678.

34   Leahy DJ, Hendrickson WA, Aukhil I, Erickson HP: Structure of a fibronectin type III domain from tenascin phased by MAD analysis of the selenomethionyl protein. Science 1992;258:987–991.

35   Adler M, Lazarus R, Dennis M, Wagner G: Solution structure of kistrin, a potent platelet aggregation inhibitor and gp IIb-IIIa antagonist. Science 1991;253:445–448.

36   Saudek V, Atkinson R, Pelton J: Three-dimensional structure of echistatin, the smallest active RGD protein. Biochemistry 1991;30:7369–7372.

37   Gan ZR, Gould RJ, Friedman PA, Polokoff MA: Echistatin: A potent platelet aggregation inhibitor from the venom of the viper, *Echis carinatus*. J Biol Chem 1988;263:19827–19832.

38   Rossi F, Billetta R, Ruggeri Z, Zanetti M: Engineered idiotypes. Immunochemical analysis of antigenized antibodies expressing a conformationally constrained Arg-Gly-Asp motif. Mol Immunol 1995;32:341–346.

39   Lanza P, Felding-Haberman B, Ruggeri ZM, Zanetti M, Billetta R: Antigenized antibodies expressing conformationally constrained RGD peptides. Studies on adhesion of human tumor cells, in preparation.

40   Shattil SJ, Hoxie JA, Cunningham M, Brass LF: Changes in the platelet membrane glycoprotein IIb.IIIa complex during platelet activation. J Biol Chem 1985;260:11107–11114.

41   Taub R, Gould RJ, Garsky VM, Ciccarone TM, Hoxie J, Friedman PA, Shattil SJ: A monoclonal antibody against the platelet fibrinogen receptor contains a sequence that mimics a receptor recognition domain in fibrinogen. J Biol Chem 1989;264:259–265.

42   Tomiyama Y, Brojer E, Ruggeri ZM, Shattil SJ, Smiltneck J, Gorski J, Kumar A, Kieber ET, Kunicki TJ: A molecular model of RGD ligands. Antibody D gene segments that direct specificity for the integrin alpha IIb beta 3. J Biol Chem 1992;267:18085–18092.

43   Tomiyama Y, Tsubakio T, Piotrowicz RS, Kurata Y, Loftus JC, Kunicki TJ: The Arg-Gly-Asp (RGD) recognition site of platelet glycoprotein IIb-IIIa on nonactivated platelets is accessible to high-affinity macromolecules. Blood 1992;79:2303–2312.

44   Niiya K, Hodson E, Bader R, Byers WV, Koziol JA, Plow EF, Ruggeri ZM: Increased surface expression of the membrane glycoprotein IIb/IIIa complex induced by platelet activation. Relationship to the binding of fibrinogen and platelet aggregation. Blood 1987;70:475–483.

45 Decker DJ, Boyle NE, Klinman NR: Predominance of nonproductive rearrangements of VH81X gene segments evidences a dependence of B cell clonal maturation on the structure of nascent H chains. J Immunol 1991;147:1406–1411.

46 Chothia C, Lesk AM, Tramontano A, Levitt M, Smith-Gill SJ, Air G, Sheriff S, Padlan EA, Davies D, Tulip WR, Colman PM, Spinelli S, Alzari PM, Poljak RJ: Conformations of immunoglobulin hypervariable regions (see comments). Nature 1989;342:877–883.

47 Ruoslahti E, Pierschbacher MD: Arg-Gly-Asp: A versatile cell recognition signal. Cell 1986;44: 517–518.

48 Plow EF, Pierschbacher MD, Ruoslahti E, Marguerie GA, Ginsberg H: The effect of Arg-Gly-Asp-containing peptides on fibrinogen and von Willebrand factor binding to platelets. Proc Natl Acad Sci USA 1985;82:8057–8061.

49 Pytela R, Pierschbacher MD, Ruoslahti E: A 125/115-kDa cell surface receptor specific for vitronectin interacts with the arginine-glycine-aspartic acid adhesion sequence derived from fibronectin. Proc Natl Acad Sci USA 1985;82:5766–5770.

50 Pytela R, Pierschbacher MD, Ginsberg MH, Plow EF, Ruoslahti E: Platelet membrane glycoprotein IIb/IIIa: Member of a family of Arg-Gly-Asp-specific adhesion receptors. Science 1986;231:1559–1562.

51 Gehlsen K, Argraves W, Pierschbacher M, Ruoslahti E: Inhibition of in vitro tumor cell invasion by Arg-Gly-Asp-containing synthetic peptides. J Cell Biol 1988;106:925–930.

52 Humphries MJ, Olden K, Yamada KM: A synthetic peptide from fibronectin inhibits experimental metastasis of murine melanoma cells. Science 1986;233:467–470.

53 Humphries MJ, Yamada KM, Olden K: Investigation of the biological effects of anti-cell adhesive synthetic peptides that inhibit experimental metastasis of B16-F10 murine melanoma cells. J Clin Invest 1988;81:782–790.

Maurizio Zanetti, The Department of Medicine and Cancer Center, University of California San Diego, 9500 Gilman Drive, La Jolla, CA 92093-0063 (USA)

Capra JD (ed): Antibody Engineering.
Chem Immunol. Basel, Karger, 1997, vol 65, pp 179–206

....

Immunogenicity of Viral Epitopes Expressed on Genetically and Enzymatically Engineered Immunoglobulins

Constantin A. Bona, Adrian Bot, Teodor-D. Brumeanu

Mount Sinai School of Medicine, Department of Microbiology, New York, N.Y., USA

The advances of peptide chemistry made possible large-scale preparation of synthetic peptides. The synthetic peptides provided a new tool for immunological studies and had a great impact on the characterization of antigenic determinants (epitopes) of protein molecules, the understanding of the molecular mechanisms of antigen recognition by T cells and the identification of microbial epitopes able to elicit a protective immune response.

The knowledge generated from studies based on the utilization of synthetic peptides opened the window for practical applications. Thus, the characterization of pathogenic self-peptides recognized by T cells which are responsible for the occurrence of autoimmune or allergic diseases represented a new avenue for immunointerventions. The characterization of microbial protective epitopes led to the development of a new generation of vaccines devoid of side effects.

The major drawback in the utilization of synthetic peptides as immunotherapeutic agents or vaccines relates to their short half-life and poor immunogenicity. The ascension of bioengineering technology allowed the construction of chimeric molecules expressing peptides chosen 'à la carte' based on their biological properties.

The basic principle of preparing chimeric molecules consists in the insertion of a 'minigene' encoding desirable peptides into the gene encoding the carrier molecule or alternatively, covalent attachment of the peptide to the carrier molecule. These approaches were successfully used to express immunogenic peptides in various microbial protein molecules. Chimeric molecules were used to evaluate the immunogenicity of microbial epitopes expressed in various carrier molecules and recognized by either B or T cells [1–5].

Studies on the immunogenicity of foreign epitopes expressed in various carrier molecules have shown that:

(a) The foreign epitopes are immunogenic and in certain cases can induce the synthesis of protective antibodies [6–8].

(b) The foreign epitopes recognized by T cells where the flanking regions derived from carrier molecules play an important role [9] in certain circumstances while in others they do not [10].

(c) Both humoral- and cellular-mediated responses are elicited subsequent to administration of chimeric molecules not only against the foreign epitope, but also against the multitude of epitopes borne by the carrier molecules [11].

These observations incited the investigators to express the foreign epitopes in *self* macromolecules. Several *self* molecules, such as the immunoglobulins [10, 12, 13], ferritin [14] or α_2-macroglobulins [15], were used as carriers to deliver foreign epitopes either genetically expressed or chemically attached. This review will focus on the delivery of foreign epitopes by immunoglobulins.

Immunoglobulins have the advantage of long half-life, ability to bind to surface receptors expressed on antigen-presenting cells (APCs) and the ability to either cross placenta or to home in tissues involved in local immunity.

Immunoglobulins as Delivery Systems

Macromolecules are currently used as platforms to deliver peptides or other biological active ligands. Among the macromolecules of various origin, the major advantage of immunoglobulins is their *self* origin. Mammals are tolerant to isotypic and cross-reactive idiotypic determinants [16]. Since the major side effects may be related to antibodies against the allotypes and individual idiotypes, such antibodies can affect the half-life by accelerating the clearance of immunoglobulins or by generation of immunocomplexes which sometimes have deleterious effects. Several approaches based on chemical, enzymatic or molecular methods were used to prepare immunoglobulins either coupled to foreign peptides or expressing foreign peptides or biologically active ligands.

Antibody-Peptide Constructs Prepared by Chemical Methods

The principle of the chemical methods consists in coupling of peptides derivatized with homobifunctional or heterobifunctional cross-linkers to immunoglobulins. For example, Wyss-Cosay et al. [13] prepared a construct by coupling cysteine-elongated tetanus toxoid peptides corresponding to amino acid residues 1273–1284 or 830–843 to derivatized monoclonal anti-CD4 or -CD8 antibodies using an N-succinimidyl 3-(2 pyridyldithioproprionate)

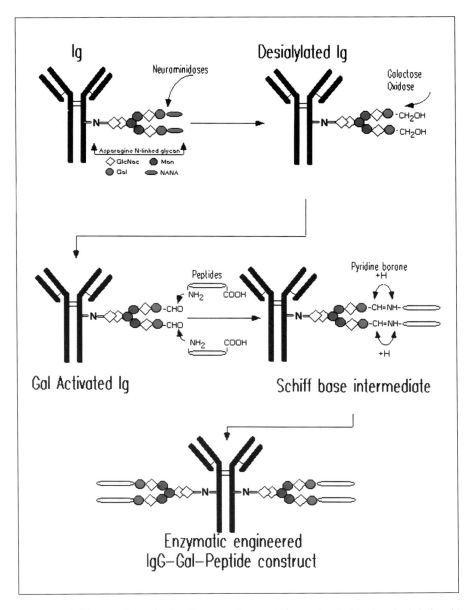

Fig. 1. Enzymatic synthesis of immunoglycopeptide constructs. The terminal sialic acid residues of the N-glycans of immunoglobulins are removed by treatment with neuraminidase from Clostridium and Arthrobacter. In the second step, the adjacent galactose residues of the desialylated immunoglobulins are oxidized with galactose oxidase, and the N-terminus of the peptides forms intermediate Schiff bases. The double bond is then reduced with pyridine borane.

(Pierce) cross-linker. It was shown that both preparations of tetanus-toxoid-derived peptides chemically coupled to anti-CD4, CD8 or CD2 monoclonal antibodies (mAbs) [13] or pigeon-cytochrome-derived peptides linked to anti-immunoglobulin, anti-class I or anti-class II MHC antigens were able to stimulate T cell clones [17]. Compared to the peptide alone, a lower molar concentration of antibody-peptide construct was required to achieve similar immune response. However, chemically engineered antibody-peptide constructs have some disadvantages:

(a) Generation of neoantigenic determinants either subsequent to binding of peptide to immunoglobulin molecule or by reactivity of some chemical groups of the cross-linker.

(b) Difficulty to predict the number of peptide molecules which can be coupled per molecule of carrier.

(c) Difficulty to predict the optimal coupling ratio between peptide and immunoglobulins in order to prevent aggregation. The aggregates are rapidly cleared from the circulation or can cause imflammatory reactions subsequent to binding to cells expressing Fc receptor (FcR).

(d) Lack of reproducibility from batch to batch with respect to the number of peptide units coupled per molecule of immunoglobulins.

Antibody-Peptide Constructs Prepared by Enzymatically-Glycosidic-Mediated Conjugation

Brumeanu et al. [18] developed a novel enzymatic procedure to couple peptides to the sugar moieties of immunoglobulins. The principle of this method illustrated in figure 1 consists in enzymatic galactose oxidation of desialylated immunoglobulins followed by covalent attachment of peptides with concurrent stabilization of imidic bonds upon mild reduction. Using this method, we prepared immunoglobulin-galactose-peptide conjugates (IGP) of various murine and human immunoglobulin isotypes. The peptides linked to immunoglobulins were derived from viral or toxoid proteins. Estimation of the efficacy of coupling of an influenza-virus-hemagglutinin-derived peptide showed an average of 11.4 peptide/IgG1 and 3.4 peptide/IgG2b molecules. Specificity of peptide attachment to the sugar moieties of immunoglobulins was demonstrated either by the ability of N-glycosidases to release the peptide from immunoglobulins or by identification of a different electrophoretic pattern of the N-linked oligosaccharides. Thus treatment of the IGPs with PGNase F able to cleave the entire N-asparagine-linked sugar completely released the coupled peptides to the galactose residues of the immunoglobulin molecule (fig. 2). This indicated that the peptides were attached exclusively to N-linked glycans of the immunoglobulin molecules [18]. This conclusion was strengthened by the results of the analysis of N-linked oligosaccharides isolated

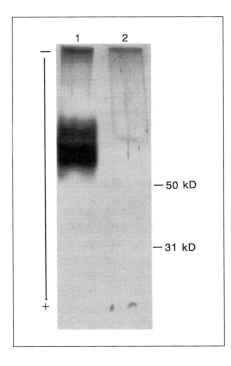

1   2

— 50 kD

— 31 kD

−

+

Fig. 2. Western blot analysis of IGP conjugate. IGP conjugate was treated or not with N-glycosidase (PGNase F) and the HA110–120 linked peptide was revealed with iodinated rabbit anti-HA110–120 peptide antibodies. Lane 1: SDS-PAGE under denaturing conditions of IGP conjugate; lane 2: SDS-PAGE under denaturing conditions of the PGNase-F-treated conjugate.

from the IGP conjugate. The molecular mass of peptidized N-oligosaccharides was significantly higher than the molecular mass of the N-linked oligosaccharides isolated from the nonconjugated immunoglobulin. By contrast, the O-linked oligosaccharides isolated from the IGP conjugates or native IgG2b molecules showed similar electrophoretic patterns (fig. 3). This indicates that the specificity of enzymatic coupling was restricted absolutely to the N-glycans.

Glycosidically engineered conjugates represent an efficient delivery system of peptides and offer several advantages:

(a) The branched architecture of the sugar moiety usually expresses four galactose acceptors per molecule of IgG and up to 20 galactose acceptors per molecule of IgM.

(b) High specificity of the coupling reaction is conferred by the enzymatic reaction.

(c) The enzymatic coupling does not alter the biological properties of the peptide attached to immunoglobulin nor the biological functions of the immunoglobulin molecule itself. The physiological properties of the peptide, the solubility of the conjugate as well as the half-life of immunoglobulin remain unaltered [18].

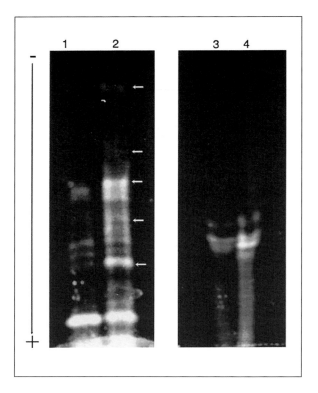

Fig. 3. Analysis of isolated N-linked oligosaccharides from the IGP conjugates using the Glyco FACE system. Lane 1: the N-glycans of the IgG2b; lane 2: N-glycans of the IgG-gal-HA conjugate; lane 3: O-glycans of the IgG2b; lane 4: O-glycans of the IgG-gal-HA conjugate.

(d) The method is rapid and the IGP conjugate can be easily purified from the reaction mixture by standard chromatographic techniques. This method allowed to prepare IGPs carrying influenza-virus-hemagglutinin- and tetanus-toxoid-derived peptides which are recognized by CD4 T cells, or viral peptides which are recognized by B cells.

Chimeric Immunoglobulins Expressing Foreign Peptides or Biologically Active Ligands

Chimeric immunoglobulins expressing foreign peptides or biologically active ligands were prepared by genetic engineering. The genetic methods that were used to prepare chimeric molecules consist in: (a) insertion or replacing complementarity-determining region (CDR) segments of V genes with oligo-

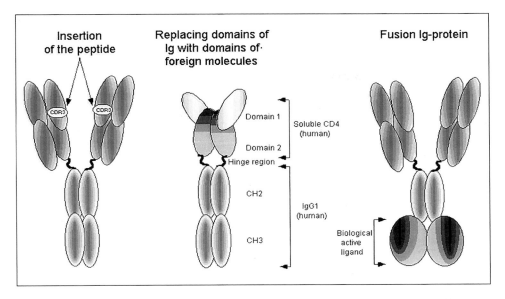

Fig. 4. Schematic representation of various types of chimeric immunoglobulin molecules prepared by genetic engineering.

nucleotides encoding foreign peptides; (b) replacing domains of heavy or light chain genes with gene segments encoding other foreign or *self* molecules and (c) the fusion of genes encoding biologically active ligands to the C_H3 domain of the heavy chain of immunoglobulins (fig. 4).

Chimeric Molecules Expressing Foreign Epitopes in the CDRs of Variable Regions of Immunoglobulin Molecules. The idea of expressing foreign epitopes in the CDRs of variable region of immunoglobulin molecules roots in the internal-image concept of Lindeman [19] and Jerne [20]. An internal image is defined by anti-idiotypic antibodies which cross-react with foreign antigens, meaning that their idiotypes represent the positive imprint of the antigens. Anti-idiotype antibodies represent a heterogeneous population of antibodies among which those carrying the internal image of antigens are also called $Ab_2\beta$. These antibodies represent a small fraction of the anti-idiotypic repertoire [21].

At the Nobel symposium in 1987 together with Ertl [22], we proposed three criteria to define an internal-image immunoglobulin:

(a) The immunochemical criterion is based on the property of $Ab2\beta$ to recognize idiotypes associated with the combining site and therefore their binding to idiotype is inhibited by antigen. However, this property which Nisonoff and Lamoyi [23] considered as important to define an internal-image immunoglobulin, cannot distinguish $Ab_2\beta$ from other anti-idiotype antibodies such as Ab$_2\gamma$ which recognize idiotypes associated with the combining site, but which does not mimick the antigen.

(b) The functional criterion is based on the ability of Ab2β to mimick the antigen and therefore to elicit an immune response similar or identical to that elicited by the nominal antigen.

(c) The structural criterion is based on sharing homologous sequences between the CDR of Ab2β and the nominal antigen. It was actually shown that Ab2βs share identical sequences with foreign antigens. Thus, Roth et al. [24] showed that an anti-idiotype monoclonal antibody (mAb) specific for GAT IdX and recognizing GAT polymer, shared the sequences Tyr-Tyr-Glu or Glu-Glu-Tyr between D segment and Glu-Tyr immunodominant epitopes of GAT synthetic terpolymer. Bailey et al. [25] showed that the injection of such Ab$_2$β in mice elicited the production of antibodies which bound to GT or GAT and expressed GAT IdX. There are other examples which showed sequence homology between the CDR of V regions and foreign antigen or allotypic antigen specificities [26, 27]. Solazzo et al. [12] first exploited the internal-image concept to insert in the CDR3 of a V$_H$ region of an autoantibody a peptide corresponding to a tandem repeat of circumsporozoite protein of *Plasmodium falciparum*. This chimeric immunoglobulin was able to elicit the production of antibodies which prevented the penetration of sporozoite into hepatocytes [28]. Similarly, Lee et al. [29] used a recombinant DNA technique to insert in the CDR3 of V$_L$ gene of the REI protein the RGD tripeptide. This sequence is involved in the binding to a number of proteins such as fibrinogen, fibronectin, vitronectin and von Willebrand's factor to integrin cell receptors such as platelet receptor αIIb β3.

The insertion of sequences encoding foreign epitopes in the CDR3 of V$_H$ or V$_L$ genes is limited to proteins capable of providing a molecular scaffold to install foreign sequences without altering the conformation of CDR3 in the way to preclude exposure of foreign sequence for binding to a cell receptor. Surface-phase exposure of the inserted peptide depends on its projection from the surface of the scaffold or on retaining the flexibility within the scaffold [29]. To overcome these constraints, Zaghouani et al. [10] chose a different strategy of genetic engineering aimed at replacing the CDR3 of the V$_H$ region with an influenza virus nucleoprotein epitope recognized by CD8+ CTL [10] or a hemagglutinin epitope recognized by CD4+ T cells [30] or the V$_3$C loop of gp120 of HIV-1 recognized by B cells [31]. This method was successfully used to replace the CDR2 with a B cell epitope (HA150–160) and CDR3 with a CD4 T cell epitope (HA110–120) of influenza virus hemagglutinin and therefore to prepare doubly chimeric immunoglobulins [32] (fig. 5).

The transfection of doubly chimeric V$_H$ genes with V$_K$ genes led to generation of transfectoma producing immunoglobulin molecules in which both B

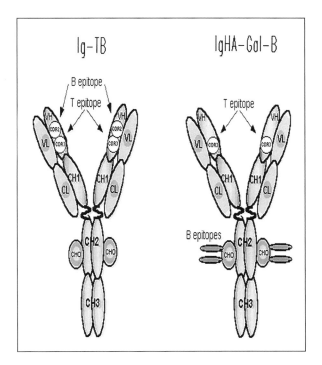

Fig. 5. Schematic representation of doubly chimeric immunoglobulin molecules prepared by genetic and enzymatic engineering.

and T cell epitopes were solvent phase exposed and retained their antigenicity. The doubly chimeric immunoglobulin molecules bound to rabbit anti-HA110–120 antibodies as well as to B2H1 mAb specific for the HA 150–160 peptide [32]. Furthermore, Brumeanu et al. [32] showed that this molecule activated the production of IL-3 by an HA110–120-specific T cell hybridoma in vitro. When this molecule was injected to mice, it was able to induce the production of antibodies specific for the HA150–160 peptide.

It therefore appears that the method of replacing the CDR with oligonucleotides encoding foreign epitopes is far superior to the method of inserting oligonucleotides in CDRs because:

(a) It allows the expression of longer peptides, particularly when they are expressed in the CDRs, which naturally exhibit a great length variability.

(b) It allows the expression of foreign epitopes in several CDRs without precluding heavy-light chain association within transfectoma and solvent phase exposure of foreign peptides, a prerequisite for their interaction with cell receptors [33].

Chimeric Immunoglobulin Prepared by Replacing V Region with Domains of Foreign Antigens. DNA recombinant technology also allowed not only the insertion of small gene segments in foreign molecules but also replacing domains of immunoglobulin molecules with domains encoded by other genes. HIV-1 through its *env* gene product, i.e gp120 protein, binds to CD4 antigen expressed on T cells. CD4 antigen functions as a viral receptor. Parenteral administration of CD4 was foreseen as an immunotherapeutic approach to prevent the penetration in, or spreading the virus in the CD4 T cells. However, the major drawback of this approach was related to the very short half-life of CD4. To overcome this, the investigators produced chimeric molecules in which CD4 domains responsible for the interaction with HIV-1 gp120 protein were fused with exons encoding the constant region of immunoglobulins [34, 35]. These molecules called immunoadhesins possess both the gp120 binding activity and HIV-blocking properties of soluble CD4 with the additional properties of immunoglobulins, notably a long half-life. Byrn et al. [35] have shown that the immunoadhesin can cause ADCC-mediated killing of HIV-1-infected cells. We have also prepared immunoadhesin molecules in our laboratry [36]. We showed that by virtue of its ability to bind to FcR, the immunoadhesin molecule favored the penetration of the HIV-1 virus in macrophages. Because of this harmful effect of the immunoadhesins, they have never been used in clinical trials. Perhaps this methodology can be successfully used for the development of other biologically active immunoglobulin chimeric molecules.

Fusion Proteins. Genetic engineering methods also allowed the preparation of chimeric immunoglobulin molecules in which certain segments of genes encoding immunoglobulin domains were fused with genes encoding other biologically active ligands such as CTLA-4 which interacts with B7.2 receptor [37], CD40 ligand which interacts with CD40, and Fas receptor (hFc Fas) which interacts with Fas ligand.

These fusion proteins represent invaluable reagents for fundamental studies and may have practical applications. Tao and Levy [38] used this methodology to produce a fusion protein in which the V_H gene encoding the idiotype of a B lymphoma was fused with the gene encoding granulocyte-macrophage colony-stimulating factor. This chimeric tumor idiotypic fusion protein elicited a strong anti-idiotype response and increased the survival of animals challenged with tumor cells.

Chimeric Immunoglobulins Expressing Peptides Recognized by T Cells

There was some information on the anti-idiotypic antibodies specific for idiotypic antibodies interacting with foreign antigens that can stimulate T cells like the antigens [39, 40]. However, there was no evidence that such antibodies represent internal image of antigens except the observation that cytotoxic T lymphocytes (CTLs) specific for hemagglutinin of reovirus type I were able to kill hybridomas producing $Ab_2\beta$ exhibiting a tetrapeptide homology with the CDR2 of the V_L region of an anti-Id antibody specific for the idiotype of anti-reovirus hemagglutinin [26].

Subsequent to Doherty and Zinkernagel's discovery of genetic restricted recognition of antigens by T cells and the demonstration that T cells recognize peptides rather than native proteins, we proposed that T cells must recognize idiopeptides derived from the processing of immunoglobulins rather than idiotypes expressed on the surface of native immunoglobulins [41].

Ensuing years proved our prediction to be correct since it was shown that (a) the CD8+ T cells recognize only short idiopeptides derived from the processing of antigens by endogenous pathway which are associated with class I antigens [42], while CD4 T cells recognize larger peptides derived from the processing of foreign antigens in the endosomal compartments [43] or endogenous pathway [44] which are associated with class II antigens. These findings correlate with others which demonstrated the isolation from MHC molecules of APCs of peptides derived from immunoglobulin molecules [45, 46].

We used chimeric immunoglobulin molecules expressing viral epitopes to study the cellular mechanisms of T cell recognition and antigen presentation of linear foreign epitopes expressed in *self* molecules. In these studies we used both genetically and enzymatically engineered chimeric immunoglobulin molecules. Genetically engineered immunoglobulins express peptides recognized by either CD8 or CD4 T cells. Ig-NP is a chimeric immunoglobulin in which the CDR3 of the V_H region was replaced with a peptide corresponding to 147–161 residues of influenza PR8 virus nucleoprotein which is recognized by CD8+ CTLs in association with K^d class I molecules [47]. Ig-HA is a chimeric immunoglobulin in which the CDR3 was replaced with a peptide corresponding to the 110–120 residues of PR8 influenza virus hemagglutinin recognized in association with I-E^d class II molecules [48]. It should be pointed out that the parental antibody used to prepare chimeric immunoglobulins lost its specificity for arsonate subsequent to replacement of CDR3 with viral peptides [10].

Enzymatically engineered immunoglobulins represent IGP conjugates in which HA110–120 was linked to the galactose residues of either IgM (i.e. IgM-gal-HA) or IgG (i.e. IgG-gal-HA). The peptides expressed by these constructs were surface exposed since rabbit polyclonal anti-HA peptide antibodies bind to Ig-HA, IgM-gal-HA or IgG-gal-HA [32].

Table 1. NP-specific CTLs lyse transfectomas expressing chimeric V_H-NP gene fail to kill target cells incubated with soluble Ig-NP molecules

Target cells	Antibody added during cytotoxic assay	Lysis by NP-specific CTL
SP2/0		–
SP2/0 infected with PR8 virus		+
SP2/0 incubated with B Lee virus		–
SP2/0 incubated with NP-K^d peptide		+
SP2/0 incubated with NP-D^d peptide		–
SP2/0 incubated with IgG$_2$b		–
SP2/0 incubated with Ig-NP		–
SP2/0 transfected with V_H NP-V_L genes		+
SP2/0 transfected with V_H gene		+
SP2/0 transfected with V_H gene	Rabbit anti-NP peptide	+
SP2/0 transfected with V_H gene	Anti-K^d mAb	–
SP2/0 transfected with V_H gene	Anti-D^d mAb	+

Immunogenicity of Peptides Expressed in Chimeric Immunoglobulins

It is well established that the peptides recognized by CD8 + T cells associated with class I MHC antigens are generated via the endogenous pathway [49]. Transfectomas bearing chimeric V_H-NP and V_K gene able to produce Ig-NP were lysed by NP-specific CTLs. Furthermore, the lysis was inhibited by the addition of anti-K^d but not anti-D^d mAb during a 4-hour cytotoxic assay. In contrast, soluble Ig-NP was unable to sensitize target cells to be lysed by NP-specific CTLs [10]. The results of these experiments are summarized in table 1. From these results several conclusions can be drawn:

(a) The generation of viral peptide, able to bind class I molecules requires endogenous expression of chimeric V_H-NP gene and the processing of chimeric V_H-NP protein in the cytosol. The peptide generated by the endogenous pathway is bound to K^d class I molecules and the complex is translocated on the surface of target cells where it is recognized by CTLs.

(b) The peptide can be generated independently of its flanking regions since a similar specific cytotoxicity was observed with PR8 virus infected or V_H-NP gene transfected target cells.

(c) Internalized soluble chimeric Ig-NP probably processed in the endosomal compartment cannot sensitize the target cells because the viral peptide is either fragmented to amino acids or if it is produced, cannot bind to class I molecules which are synthesized in the endoplasmic reticulum and then exported to the membranes.

Peters et al. [50] showed that the class I peptide pathway does not intersect the class II peptide pathway. These conclusions were supported by in vivo experiments. No NP-specific CTL activity was detected in BALB/c mice immunized with soluble Ig-NP in saline or FCA in newborn mice immunized three times at 2-day intervals with 1 mg soluble Ig-NP, nor in the progeny born from mothers injected with 1 mg Ig-NP on day 16, 18 and 20 of the pregnancy. By contrast, the NP-specific CTL precursors were expanded subsequent to injection with irradiated transfectomas expressing the chimeric V_H-NP gene [51]. Our results strongly suggest that the internal-image immunoglobulins are unable to stimulate the CD8 T cells. The cells are stimulated only if the internal-image V gene products are processed in the cytosol.

Cao et al. [52] reported a sequence homology between A/Japan influenza-virus-hemagglutinin-derived peptide recognized by CTL and peptide of the V_H gene belonging to the V_H7183 family. They also showed that hemagglutinin-specific CTL successfully lysed hybridomas expressing this V_H7183 gene. This observation suggests that peptides derived from the processing of V domains of immunoglobulins may represent a positive selection force of T cells [53] because idiopeptides can mimic the foreign peptides. In contrast to the soluble Ig-NP, which cannot generate the viral peptide able to be recognized by CD8 + T cells in association with class I MHC molecules, both genetically or enzymatically engineered immunoglobulins expressing viral epitopes can stimulate the CD4 T cells. Our studies demonstrated that Ig-HA molecules are internalized mainly via FcR since the pretreatment of APC with anti-FcR mAb drastically inhibits the activation of T cells.

Internalized Ig-HA is processed in the endosomal compartment since the chloroquine treatment or prefixation of APC with paraformaldehyde inhibited the activation of T cells. HA110–120 peptide released from the CDR3 region of V_H binds to I-E^d and the complex is translocated on the surface where it is recognized by T cells. The activation of T cells was inhibited by anti I-E^d but not anti I-A^d mAbs [30]. The results are summarized in table 2.

A comparative study of the potency in presentation of HA110–120 peptide following the processing of Ig-HA by various APCs showed that the maximal activation of 2×10^4 T cell hybridoma by 15 µg/ml Ig-HA required either 3×10^3 B cells or 10^4 dendritic cells or 10^6 spleen cells as APCs [54]. Thus, it appears that in vivo, various professional APC exhibit various degrees of efficacy in processing chimeric immunoglobulin molecules.

It should be pointed out that chimeric immunoglobulins stimulate not only the activation of T hybridoma cells expressing a TCR specific for HA110–120 peptide I-E^d complex but also resting CD4 or CD8 T cells expressing TCR genes [55]. T cells of the TCR-HA transgenic mice express the $V\beta8.2$ and $V\alpha4$ genes of the 14.3d TCR of 14-3-1 TcH which recognizes HA110–120 epitope

Table 2. Effect of chloroquine, paraformaldehyde fixation and anti-class II mAbs on the activation of T cell hybridoma by genetically and enzymatically engineered immunoglobulins expressing HA110–120 peptide

Treatment of APC	Activation of T cells by				
	HA110–120 peptide	PR8 virus	Ig-HA	IgM-gal-HA	IgG-gal-HA
Nil	+	+	+	+	+
Chloroquine	+	–	–	–	+
Paraformaldehyde	+	–	–	–	+
Anti-FcR mAb	+	+/–	–	+	–
Anti-I-E^d mAb	–	–	–	–	–
Anti-I-A^d mAb	+	+	+	+	+

of the influenza virus hemagglutinin [56]. T cell hybridoma specific for HA110–120 peptide was stimulated by both Ig-HA and IgG-gal-HA constructs.

Recently, Bentley et al. [57] determined the crystal structure of extracellular portion of Vβ8.2 chains of 14.3d T cell hybridoma. Peptide loops analogous to CDRs of the V region of immunoglobulins are disposed on the surface, and they contribute to the binding of the peptide, MHC molecules and superantigens.

The CDR1 is stabilized by the Gln25 residue which, by its amide group, forms four hydrogen bonds to its main chain atoms. The CDR2 is stabilized by the side chain of Ser49 which forms hydrogen bonds with the main chain atoms. Both CDRs display restricted conformational heterogeneity suggesting that they primarily contact the I-E molecule. CDR1 and CDR2 are juxtaposed to CDR3 which is supposed to interact with the peptide exposed to the solvent in the I-E groove. Vβ8.2 has an additional CDR (HV4) comprising residues 64–75 which fold toward CDR1 and CDR2. This additional CDR may contribute to the binding of TcR to the MHC-peptide complex. The Vβ domain is completely solvent phase exposed and comprises net positively charged residues. These residues cannot be neutralized by contact with a neighboring Cβ domain, but rather form a salt bridge with residues of Cα in heterodimer molecules.

The data depicted in figure 6 show the proliferation of T cells from nonimmunized transgenic mice upon in vitro incubation with various carriers expressing HA110–120 peptide. However, Brumeanu et al. [58] showed a strong proliferative response of T cells from normal mice primed with chimeric immunoglobulin injected in FCA (table 3).

Table 3. Proliferative response of T cells isolated from mice immunized with chimeric immunoglobulin expressing a viral CD4 immunodominant epitope HA 110-120

Mice immunized with 100 µg of antigen in FCA	In vitro proliferative response upon exposure to						
	ConA	PPD	HA110–120	HA150–159	PR8 virus	BHA	B Lee virus
Ig-HA	+	+	+	–	+	+	–
IgG-gal-HA	+	+	+	–	+	+	–
IgG₂b	+	+	–	–	–	–	–
IgM-gal-HA	+	+	+	–	+	+	–
IgM	+	+	–	–	–	–	–
HA110–120 peptide	+	+	+	–	+	+	–

In aggregate, these data demonstrate that both genetically and enzymatically engineered immunoglobulins which express HA110–120 peptide stimulate HA110–120 specific T hybridoma cells like BHA, PR8 virus as well as the proliferation of resting lymphocytes from transgenic mice. These constructs can also prime the precursors of HA110–120 specific T cells in normal animals.

Immunopotency of HA110–120 Peptide Expressed in Chimeric Molecules
Brumeanu et al. [58] evaluated the potency of HA110–120 peptide expressed in various carriers by measuring the activation of T cell hybridoma. The amount of peptide for each carrier was normalized based on molar equivalents of HA110–120 peptide for each carrier, precisely 1 nM HA110–120 peptide carried by 0.5 nM Ig-HA, 0.35 nM IgG-gal-HA or 0.3 nM BHA. At 1 nM peptide concentration, IgG-gal-HA was as active as BHA, one log more active than Ig-HA and four logs more active than the synthetic peptide. These results show that the presentation of a viral peptide by various carriers is dependent on the nature of the carrier as well as on the cellular receptors which contribute to their internalization in APCs. These differences may be related to the kinetics of generation of peptide from various carriers or to molecular mechanisms involved in the presentation of peptide to I-E molecules.

Kinetics of Generation and Persistence of Viral Peptide Expressed in Various Carrier Molecules
There are several parameters used to determine the immunogenicity of a peptide recognized by T cells. First, it is important to establish whether the peptide can stimulate the T cells when it is presented to APCs in different flanking regions. Secondly, whether it is related to the kinetics of the generation of peptide, namely whether it is generated from natural molecular environments

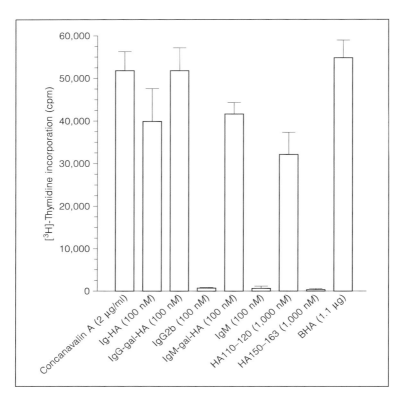

Fig. 6. In vitro proliferation of T cells obtained from nonimmunized TcR-HA transgenic mice upon incubation with various constructs.

or in a different molecular context. Thirdly, whether it is related to the persistence of peptide generated in various molecular contexts on the surface of APCs.

Taking advantage of chimeric immunoglobulins expressing viral peptides, Bot et al. [59] addressed these fundamental questions. To investigate the kinetics of generation and persistence of HA110–120 peptide expressed in various carriers, Bot et al. [59] studied the activation of a HA110–120-specific T cell hybridoma transfected with *Escherichia coli* LacZ gene fused to the IL-2 gene. The occupancy of TCR of this hybridoma is followed by a rapid expression of the reporter gene and the synthesis of β-galactosidase. The results of this study show that the generation of HA110–120 peptide is dependent of the nature of the carriers. The maximal activation of TcH required 4 h for the synthetic peptide and 8 h for Ig-HA, BHA and 12 h for PR8 virus. Furthermore, Bot et al. [59] studied the effect of antibodies specific for HA110–120 peptide or anti-HA antibodies on the presentation of the peptide by various carriers. In contrast to Simitsek et al. [60], we found a dose-dependent inhibition of the activation of T cells when the synthetic peptide or Ig-HA was preincubated with polyclonal rabbit anti-HA110–120 antibodies or when BHA and PR8 virus were preincubated with PY102 a hemagglubinin-neutralizing mAb (fig. 7).

While the inhibitory effect of anti-HA110–120 antibody on synthetic peptide presentation in context of MHC class II can be explained by protection of peptide bound to the combining site of the antibody, the inhibitory effect of antibodies on the presentation of peptide derived from the processing of Ig-HA, BHA or PR8 suggests that the antibodies either prevent the internalization or alternatively direct the complex to the endosomal compartment which precludes the generation of peptide or modifies the antigen processing in the MHC class II endosomal compartment. Using iodinated Ig-HA or PR8 virus, we observed that while PR8 virus is rapidly internalized (30 min) via viral receptor, the internalization of Ig-HA via FcR or fluid-phase pinocytosis reached the peak at 3 h after the pulse. However, while the preincubation of Ig-HA with anti-HA110–120 antibodies showed a slight effect on the internalization, it practically ablated the ability to activate T cells at equivalent concentration. This suggests that after internalization, the Ig-HA-anti-HA complex is either directed to a lysosomal compartment where fragmentation precludes the generation of HA110–120 peptide or alternatively, the peptides generated by processing of rabbit antibodies compete with HA110–120 peptide for binding to empty I-E^d molecules in class II vesicles. In contrast, preincubation of PR8 virus with PY201 anti-HA mAb prevented the efficient uptake of the PR8 virus by the 2PK3 cells. These results show a negative modulation of the presentation of an immunodominant peptide recognized by CD4 T cells, either by antibodies specific for peptide or its carrier.

Lanzavecchia et al. [61] showed that under physiological conditions, the high-affinity peptides are bound irreversibly to class II molecules. We investigated the persistence of HA110–120 peptide subsequent to pulsing of APCs with various carriers for a sufficient time interval as determined from the T cell activation assay. After pulsing, the cells were extensively washed and incubated for various time intervals before coincubation for 4 h with TcH cells. A drastic decrease in activation of TcH cells by 2PK3 cells incubated 60 h after pulsing with various antigens was observed (fig. 8). This suggests that the persistence on MHC class II molecules of the HA110–120 peptide generated from various carriers corresponds to the intrinsic half-life of class II molecules and that it is independent of the nature of the carrier.

Antigen	12 h/no Abs	12 h/Abs

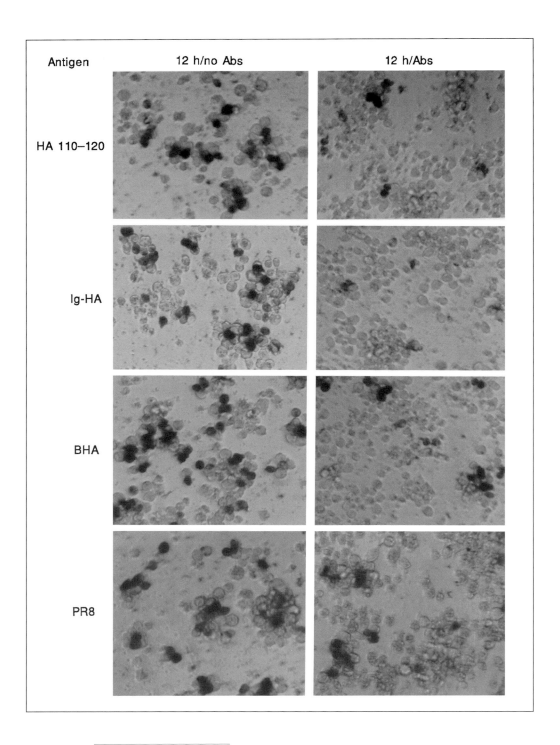

HA 110–120

Ig-HA

BHA

PR8

Molecular Mechanisms Responsible for the Presentation of Viral Epitopes Expressed in Chimeric Immunoglobulin Molecules

It is well known that the antigen presentation by class II molecules represents a complex cellular and molecular event consisting in the internalization of antigen in the endosomal compartment, enzymatic processing leading to the generation of peptides, binding of peptide to empty class II molecules and the translocation of the complex on the surface of APCs [62–64]. The data presented above clearly show that influenza-virus-hemagglutinin-derived peptide expressed in genetically or enzymatically engineered immunoglobulins stimulates the activation or proliferation of CD4 T cells specific for the peptide. These observations strongly suggest that APCs generate a peptide similar or identical with the peptide generated from the processing of viral hemagglutinin and that similar mechanisms are responsible for the generation of peptide from viral protein and self immunoglobulins expressing the peptide. However, our studies clearly demonstrate that there are subtle differences between the mechanisms of presentation of HA110–120 peptide by viral protein, genetically engineered Ig-HA and the enzymatically engineered IgG-gal-HA.

Molecular Mechanisms Responsible for Efficient Presentation of HA110–120 Peptide by Ig-HA. Study of the activation of HA110–120 specific T cell hybridoma by PR8 virus, BHA, Ig-HA and the synthetic peptide showed that Ig-HA was the most efficient when it was presented by various APC [30]. The internalization of both Ig-HA and BHA is mediated via cellular receptors since the activation of T cells by Ig-HA was inhibited by preincubation of APCs with anti-FcR [30] and that of BHA with anti-HA mAbs [59]. This indicates that while Ig-HA is mainly internalized via FcR, the virus and BHA were internalized via sialoprotein receptor.

The generation of HA110–120 from both viral and chimeric self immunoglobulins takes place in the endosomal compartment since pretreatment with chloroquine or prefixation with formaldehyde of APCs precludes the activation of T cell hybridoma. The isolation and sequencing of peptides isolated from I-E molecules of APCs pulsed with HA110–120 synthetic peptide, PR8 virus or Ig-HA showed peptides with identical structure. From Ig-HA-pulsed APCs Brumeanu et al. [65] also isolated an octamer corresponding to a truncated peptide [65]. In contrast, from I-E molecules of APCs pulsed with IgM-gal-HA conjugate, we isolated only truncated peptides, one corresponding to an

Fig. 7. Effect of preincubation of antigen with anti-HA antibodies. 14-3-1 T cell hybridoma was incubated for 12 h with 2PK3 pulsed with HA110–120 or Ig-HA alone or preincubated with rabbit anti-HA110–120 peptide antibodies or with BHA or PR8 virus preincubated with anti-HA mAbs. After incubation the cells were washed, fixed and incubated with X-Gal substrate.

Antigen	12 h	60 h	84 h

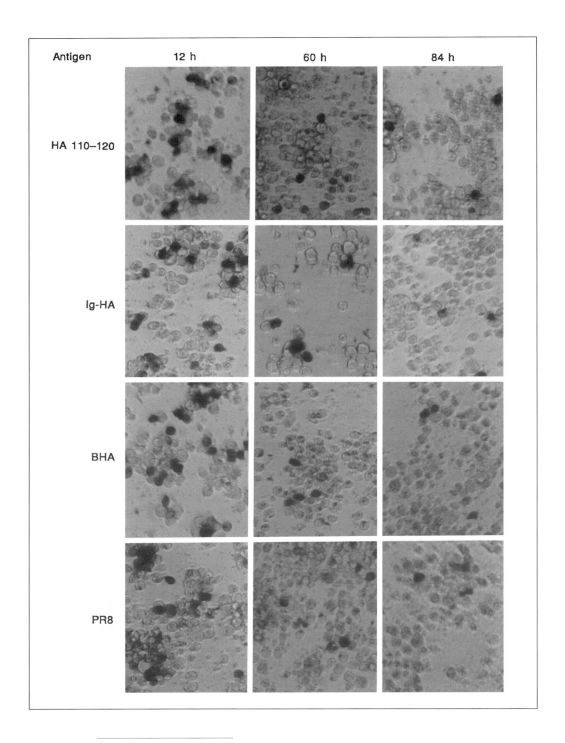

HA 110–120

Ig-HA

BHA

PR8

Table 4. Structure of peptides isolated from I-E^d molecules of 2PK3 cells pulsed with various antigens

Antigens[1]	Structure of peptides recovered from I-E^d molecules
HA110–120	SFERFEIFPKE
IgM-gal-HA	SFERFEIFP
	SFERFEIF
Ig-HA	SFERFEIFPKE
	SFERFEIF
PR8 virus	SFERFEIFPKE

[1] 10^9 2PK3 cells were pulsed with 30 mg of the antigen.

octamer and another to a nonamer (table 4). While the nonamer peptide was able to activate T cell hybridoma the octamer was not [58]. These data indicate that the peptides derived from the processing of different forms of external antigen, such as synthetic peptide, free virus, chimeric immunoglobulin or IGP conjugate, have identical or very similar structure. Thus, in spite of various degrees of fragmentation depending on the nature of the carrier, peptides of different sizes are efficiently sorted by class II molecules in endosomes. It also clearly appears that truncated peptides can be generated from both Ig-HA in which the peptide is expressed in the CDR3 as well as from IgM-gal-HA where the peptide is solvent phase exposed.

Mechanisms Responsible for the Presentation of HA110–120 Peptide by IgG-gal-HA. By studying the activation of HA110–120-specific T cell hybridomas by IgG-gal-HA, we observed that this can be achieved either using chloroquine-pretreated or formaldehyde-prefixed APCs. These observations suggest that IgG-gal-HA activates the T cell hybridoma by circumventing the processing by APCs. Taken together, these observations suggest that IgG-gal-HA bound to FcR and is able to present the HA110–120 peptide to neighboring class II molecules with concurrent recognition by T cells (fig. 9). However, this model [66] cannot exclude that subsequent to binding of IgG-gal-HA construct to FcR of APCs, the peptide may be released by proteases from serum added to the culture medium used to incubate the APCs with T

Fig. 8. Persistence of HA110–120 peptide on surface of APCs. 2PK3 cells were pulsed for 1 h with HA110–120 synthetic peptide, 5 h with Ig-HA or BHA and 8 h with PR8 virus. The cells were washed and cultured for 12, 60 or 72 h. After culture, the cells were fixed, incubated for 4 h with 14-3-1 T cell hybridomas and then incubated with X-gal substrate.

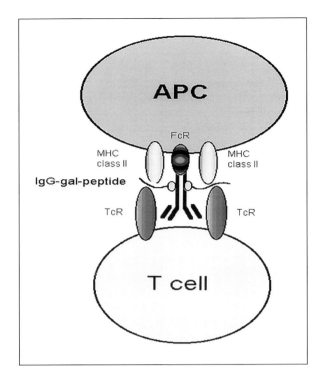

Fig. 9. Model of MHC class II presentation to T cells of HA110–120 peptide by IgG-gal-HA construct. The IgG-gal-HA conjugate is bound by the FcR on the surface of the APC and the peptide assembled on the galactose residues associates with both MHC class II and TcR molecules. Theoretically, one molecule of IgG-gal-HA may engage in the inter-action two MHC molecules and two TcR molecules.

cell hybridoma or by ectopeptidases associated with the membrane of APCs [67]. To address these questions, we took advantage of a HA110–120-specific T cell hybridoma transfected with a plasmid-expressing IL-2 gene fused with LacZ gene. The incubation of this T cell hybridoma with APCs and HA110–120 peptide rapidly activates the IL-2 gene and the activation can be measured by the synthesis of β-galactosidase [59]. This system allowed to determine:

(a) That the activation of HA110–120 specific T cells by APCs incubated with IgG-gal-HA took place in FCS-free medium.

(b) That the activation of HA110–120-specific T cells took place in the presence of inhibitors of cysteine proteases, serine proteases known to be associated with the membrane of antigen-presenting B cells [66].

(c) That paraformaldehyde-fixed B Long melanoma cells transfected with FcR do not facilitate the cleavage of HA110–120 peptide from the IgG-gal-

Table 5. Activation of T cells by IgG-gal-HA does not require antigen processing

Substances used for the incubation[1]	Activation of HA110–120 specific T cell hybridoma
Nil	+
Chloroquine	+
Formaldehyde	+
Anti-FcR mAb	–
Anti-I-E^d mAb	–
FCS-free medium	+
Aprotinin	+
Phenylmethylsulfonyl fluoride	+
N-α-*p*-tosyl-*l*-tyrosine chloromethyl ketone	+
p-Chloromercurilbenzoic acid	+

[1] APCs were exposed to IgG-gal-HA and incubated with various substances.

HA construct. Furthermore, we have shown that fixed B Long cells expressing FcR and H-2^b class I antigen incubated with 2PK3 cells were able to activate HA110–120-specific T cell hybridoma. Taken together, these results clearly demonstrate that IgG-gal-HA can present the HA110–120 peptide to T cells without processing (table 5).

Therefore, it appears that the mechanism of presentation of the peptides by IGP conjugates is different from the presentation of peptide by engineered Ig-HA which is internalized via FcR or by enzymatically engineered IgM-gal-HA that is internalized by fluid-phase pinocytosis. The fact that a peptide enzymatically assembled on the sugar moieties of immunoglobulins can circumvent antigen processing suggests the possibility of using the enzymatic method to generate constructs able to deliver protective epitopes or other biologically active ligands. Such constructs can be used for neonatal immunization because it was shown that B cells from young animals are less efficient in processing the antigens than the B cells from adult animals.

Conclusions

During the last decade, much hope was put on the use of the internal image of anti-idiotypic antibodies as molecules mimicking foreign antigens, biologically active ligands or drugs. However, the difficulties encountered in obtaining these reagents and distinguishing internal-image immunoglobulins from other anti-idiotypes have severely limited the utilization of anti-idiotypes as vaccines. The advance of molecular technology allowed the construction of chimeric immunoglobulins expressing parasitic [12] viral [16] or other biologically active peptide [29].

One of the first approaches was to add a linear parasitic epitope to a CDR3 segment of a V_H region of a hybridoma producing an autoantibody specific for thyroglobulin [12]. It was claimed that this antibody prepared 'à la carte' represents a genuine internal image antibody [28]. However, one may ask which is the significance to the internal image concept and to practical utilization of these chimeric molecules as vaccines, by expressing microbial linear epitopes in immunoglobulins.

Poljak's group succeeded in resolving the structure of an Fab fragment of an antibody specific for lysozyme [68] as well as of an antibody specific for lysozyme and an anti-idiotypic antibody carrying the internal image by X-ray crystallography [69]. These studies demonstrated that more than 15 residues belonging to almost all CDRs of the V_H and V_L domains participate to the binding of antibody to lysozyme and of anti-idiotype antibody to anti-lysozyme antibody. These observations demonstrate that the insertion of a linear epitope in a CDR of the V_H region of immunoglobulin cannot represent a valuable approach to prepare genuine internal-image antibody since $Ab_2\beta$ mimics the antigen by three-dimensional structure rather than linear epitopes [69]. This conclusion is strongly supported by the observations of Lanza et al. [70]. These investigators inserted in the CDR3 of the V_H region of an autoantibody, linear sequences corresponding to CD4 domains responsible for the binding to the HIV-1 gp120 protein. While this internal-image CD4 antibody injected in rabbits induced the production of anti-CD4 antibodies [70], injection to monkeys never elicited the production of neutralizing HIV-1 antibodies [Kennedy, pers. commun.] as should be expected from an internal-image immunoglobulin.

In contrast, either the expression of a linear viral epitope recognized by T cells in a CDR region of a V gene or by attaching it to the sugar moiety may represent a useful tool for study of molecular basis of antigen presentation by various carriers [10, 30, 58]. These chimeric immunoglobulin molecules expressing foreign epitopes may represent a new approach to the development of T cell vaccines. The approach is based on the fact that while B cells generally recognize sequential epitopes, T cells recognize linear peptides derived from the fragmentation of proteins either in the cytosolic or the endosomal compartment. Thus, we believe that the utilization of self immunoglobulins as delivery systems of linear epitopes recognized by T cells or other biologically active ligands interacting with cell receptors shall see potential applications in fundamental studies and in the development of immunotherapeutic agents.

References

1   Jennings PA, Bills M, Irving DD, Matlick JS: Fimbriae of *Bacteroides nodosus*: Protein engineering of the structural subunit for the production of an exogenous peptide. Protein Eng 1980;2:365–369.
2   Valenzuela P, Coit D, Medina-Selby MA, Kuo CH, VanNest G, Burke RL, Bull P, Urdea MS, Graves PV: Antigen engineering in yeast: Synthesis and assembly of hybrid hepatitis B surface antigen-Herpes simplex 1 gD particle. Biotechnology 1985;3:323–326.
3   Rutgers T, Gordon D, Gethoye A-M, Hollingdale M, Hockmeyer W, Rosenberg M, DeWilde M: Hepatitis B surface antigen as carrier matrix for the repetitive epitope of the circumsporozoite protein of *Plasmodium falciparum*. Biotechnology 1988;6:1065–1069.
4   Li S, Schulman JL, Moran T, Bona C, Palese P: Influenza A virus transfectants with chimeric hemagglutinins containing epitopes from different subtypes. J Virol 1992;66:399–404.
5   Leclerc C (ed): Immunogenicity of foreign epitopes expressed in chimeric molecules. Int Rev Immunol 1994;11:103–178.
6   Clarke BE, Newton SE, Caroll AR, Francis MJ, Appleyard G, Syred AD, Highfield PE, Rowlands DJ, Brown F: Improved immunogenicity of a peptide epitope after fusion to hepatitis B core protein. Nature 1987;330:381–384.
7   Evans DJ, McKeating J, Meredith JM, Burke KL, Katrak K, John H, Ferguson M, Minor PD, Weiss RA, Almond JW: An engineered poliovirus chimera elicits broadly reactive HIV-1 neutralizing antibodies. Nature 1989;339:385–388.
8   Burke KL, Dunn G, Ferguson FM, Minor PD, Almond JW: Antigen chimeras of poliovirus as potential new vaccines. Nature 1988;332:81–82.
9   Del Val M, Schlicht H-J, Ruppert T, Reddehase MJ, Koszinowski UH: Efficient processing of an antigen sequence for presentation by MHC class I molecules depends on its neighboring residues in the protein. Cell 1991;66:1145–1153.
10 Zaghouani H, Krystal M, Kuzu H, Moran T, Shah H, Kuzu Y, Schulman J, Bona C: Cells expressing a heavy chain immunoglobulin gene carrying a viral T cell epitope are lysed by specific cytolytic T cells. J Immunol 1992;148:3604–3609.
11 Newton SMC, Jacob CO, Stocker BAD: Immune response to cholera toxin epitope inserted in Salmonella flagellin. Science 1989;147:70–72.
12 Sollazzo M, Billeta R, Zanetti M: Expression of an exogenous peptide epitope genetically engineered in the variable domain of an immunoglobulin: Implications for antibody and peptide folding. Bioengineering 1990;4:215–220.
13 Wyss-Coray T, Brander C, Bettens F, Mijic D, Pichler WJ: Use of antibody-peptide constructs to direct antigenic peptides to T cells: evidence for T cell processing and presentation. Cell Immunol 1992;139:268–273.
14 McCoy KL, Noone M, Inman JK, Strutzman C: Exogenous antigens internalized through transferrin receptors activate CD4 + T cells. J Immunol 1993;150:1191–1204.
15 Mitsuda S, Nakagawa T, Nakazato H, Ikai A: Receptor-linked antigen delivery system. Biochem Biophys Res Comm 1995;216:393–405.
16 Bona C, Cazenave PA (eds): Lymphocytic Regulation by Antibodies. New York, Wiley InterScience, 1981.
17 Casten LA, Kaumaya P, Pierce SK: Enhanced T cell responses to antigen peptides targeted to B cell surface Ig Ia or class I molecules. J Exp Medods 1988;168:171–180.
18 Brumeanu TD, Dehazya P, Wolf I, Bona CA: Enzymatically mediated, glycosidic conjugation of immunoglobulins with viral epitopes. J Immunol Methods 1995;183:185–197.
19 Lindemann J: Homobodies: Do they exist? Ann Immunol Inst Pasteur 1973;130C:311–318.
20 Jerne NK: Towards a network theory of the immune response. Ann Immunol (Paris) 1974;125C: 373–389.
21 Bona C, Victor-Kobrin C, Manheimer A, Legrain P, Buttin G, Yancopoulos G, Alt F: in Pendura A, Doria G, Damarco F, Bargelli A (eds): Genetic and Molecular Aspects of Regulatory Idiotypes in Monoclonal Antibodies 84: Biological and Clinical Applications. Milano, Kurtis, 1985, pp 55–75.
22 Ertl HCJ, Bona C: Criteria to define anti-idiotypic antibodies carrying the internal image of an antigen. Vaccine 1988;6:80–84.

---

23 Nisonoff A, Lamoyi E: Implications of the presence of an internal image of the antigen in anti-idiotypic antibodies: Possible application to vaccine production. Clin Immunol Immunopathol 1981;21:397–406.

24 Roth C, Rocca-Serra J, Somme G, Fougereau M, Theze J: Gene repertoire of the anti-poly (Glu^{60}, Ala^{30}, Fyr^{10}) GAT Immune response: Comparison of VH, VK and D regions used by anti-GAT antibodies and monoclonal antibodies produced after antiidiotypic immunization. Proc Natl Acad Sci USA 1985;82:4788–4792.

25 Bailey NC, Fidanza V, Mayer R, Mazza G, Fougereau M, Bona C: Activation of clones producing self-reactive antibodies by foreign antigens and anti-idiotype antibody carrying the internal image of the antigen. J Clin Invest 1989;84:744–756.

26 Bruck C, Co SM, Slaoui M, Gaulton GN, Smith T, Fields BN, Mullins JI, Green MI: Nucleic acid sequence of an internal image-bearing monoclonal anti-idiotype and its comparison to the sequence of the external antigen. Proc Natl Acad Sci USA 1986;83:6578–6582.

27 Van Cleave WH, Naeve CW, Metzger DW: Do antibodies recognize amino acid side chains of protein antigens independently of carbon backbone. J Exp Med 1988;167:1841–1848.

28 Biletta R, Hollingdale MR, Zanetti M: Immunogenicity of an engineered internal antibody. Proc Natl Acad Sci USA 1991;88:4713–4717.

29 Lee G, Chan W, Hurle MR, DesJarlais RL, Watson F, Sathe GM, Wetzel R: Strong inhibition of fibrinogen binding to platelet receptor—αIIβ3 by RAD sequences installed into a presentation scaffold. Protein Eng 1993;6:745–754.

30 Zaghouani H, Steinman R, Nonacs R, Shah H, Gerhard W, Bona C: Presentation of a viral T cell epitope expressed in the CDR3 region of a self immunoglobulin molecule. Science 1993;259:224–227.

31 Zaghouani H, Anderson SA, Sperber KE, Daian C, Kennedy RC, Mayer L, Bona C: Induction of antibodies to the human immunodeficiency virus type I by immunization of baboons with immunoglobulin molecules carrying the principal neutralizing determinant of the envelope protein. Proc Natl Acad Sci USA 1995;92:631–635.

32 Brumeanu T-D, Bot A, Bona CA, Dehazya P, Wolf I, Zaghouani H: Engineering of doubly anti-genized immunoglobulins expressing T and B viral epitopes. Immunotechnology 1996;2:85–95.

33 Zaghouani H, Kuzu Y, Kuzu H, Mann N, Daian C, Bona C: Engineered immunoglobulin molecules as vehicles for T cell epitopes. Int Rev Immunol 1993;10:265–278.

34 Traunecker A, Schneider J, Kiefer A, Karjalainen K: High efficient neutralization of HIV with recombinant CD4-immunoglobulin molecules. Nature 1989;339:68–70.

35 Byrn RA, Mordenti J, Lucas C, Smith D, Marsters SA, Johnson JJ, Cossum P, Chamow SM, Wurm FM, Gregory T, Groopman JE, Capron DJ: Biological properties of a CD4 immunoadhesin. Nature 1990;344:667–669.

36 Dinca L, Zaghouani H, Brumeanu T, Mayer L, Sperber K, Bona C, Fidanza V: Effect of immu-noadhesin on infection of human T cells and monocytes by HIV-1. Autoimmunity 1992;(suppl): 49.

37 Linsley PS, Brady W, Urnes M, Grosmaire L, Damle NK, Ledbetter JA: CTLA-4 is a second receptor for the B cell activation antigen B. J Exp Med 1991;174:561.

38 Tao M-H, Levy R: Idiotype/granulocyte macrophage colony-stimulating factor fusion protein as a vaccine for B-cell lymphoma. Nature 1993;362:755–758.

39 Kaufmann SHE, Eichmann K, Muller I, Wrazel LJ: Vaccination against the intracellular bacterium *Listeria monocytogenes* with a clonotypic antiserum. J Immunol 1985;134:4123–4127.

40 Rees A, Scoging A, Dobson N, Praputpittaya K, Young D, Ivanyi J, Lamb JR: T cell activation by anti-idiotypic antibody: Mechanism of interaction with antigen-reactive T cells. Eur J Immunol 1987;17:197–201.

41 Bona C: Antiidiotypes; in Bona C (ed): Biological Applicatons of Antiidiotypes. Boca Raton, CRC Press, 1988, vol 1, pp 1–13.

42 Chakarabarti D, Ghosh SK: Induction of syngeneic cytotoxic T lymphocytes against a B cell tumor III MHC class I restricted CTL recognizes the processed form of idiotype. Cell Immunol 1992;144: 455–464.

43 Pride MW, Thakur A, Thanavala Y: Mimicry of the 'a' determinant of hepatitis B surface antigen by anti-idiotype antibody. J Exp Med 1993;177:127–134.

44   Weiss S, Bogen B: MHC class II-restricted presentation of intracellular antigen. Cell 1991;64:76–776.
45   Rudensky AY, Preston-Hurlbust P, Hong S-C, Beslow A, Janeway CA: Sequence analysis of peptide bound to MHC class II molecules. Nature 1991;353:622–627.
46   Chicz RM, Urban RG, Lane WS, Gorga JC, Stern LJ, Vignali DAA, Strominger JL: Predominant naturally processed peptides bound to HLA-DR1 are derived from MHC-related molecules and are heterogeneous in size. Nature 1992;358:764–768.
47   Taylor PM, Askonas BA: Influenza nucleoprotein-specific cytotoxic T-cell clones are protective in vivo. Immunology 1986;58:417–420.
48   Haberman AM, Moller C, McCreedy D, Gerhard WV: A large degree of functional diversity exists among helper T cells specific for the same antigenic site of influenza hemagglutinin. J Immunol 1990;145:3087–3094.
49   Townsend ARM, McMichael AJ, Carter NP, Huddleston JA, Brownlee GG: Cytotoxic T cell recognition of the influenza nucleoprotein and hemagglutinin expressed in transfected mouse L cells. Cell 1984;39:13.
50   Peters PJ, Neefjes JJ, Oorschot V, Ploegh HL, Genze HJ: Segregation of MHC class II molecules from MHC class I molecules in the Golgi complex for transport to lysosomal compartments. Nature 1991;349:669–675.
51   Kuzu Y, Kuzu H, Zaghouani H, Bona C: Priming of cytolytic T lymphocytes at various stages of ontogeny with transfectoma cells expressing a chimeric heavy chain gene bearing an influenza virus nucleoprotein peptide. Intern Immunol 1993;5:1301–1307.
52   Cao W, Myers-Powell BA, Braciale TJ: Recognition of an immunoglobulin VH epitope by influenza virus-specific class I major histocompatibility complex-restricted cytolytic T lymphocytes. J Exp Med 1994;179:195–202.
53   Kouriliski P, Claveric JM: The peptide self-model: A hypothesis on the molecular nature of immunological self. Ann Inst Pasteur 1986;137D:3–12.
54   Zaghouani H, Kuzu Y, Kuzu H, Brumeanu TD, Swiggard WJ, Steinman RM, Bona CA: Contrasting efficacy of presentation by major histocompatibility complex class I and class II products when peptides are administered within a common protein carrier, self immunoglobulin. Eur J Immunol 1993;23:2746–2750.
55   Weber S, Trannecher A, Oliveri F, Gerhard W, Karjalainen K: Specific low affinity recognition of MHC complex plus peptide by soluble T cell receptor. Nature 1992;356:793–796.
56   Kirberg J, Baron A, Jakob S, Rolink A, Karjalainen K, von Boehmer H: Thymic selection of CD8 + single positive cells with class II MHC-restricted receptor. J Exp Med 1994;180:25–34.
57   Bentley GA, Boulot G, Karjalainen K, Mariuzza R: Crystal structure of the β chain of a T cell antigen receptor. Science 1995;267:1984–1987.
58   Brumeanu TD, Casares S, Harris PE, Dehazya P, Wolf I, von Boehmer H, Bona CA: Immunopotency of a viral peptide assembled on the carbohydrate moieties of self immunoglobulins. Nature/Biotechnology 1996;14:722–725.
59   Bot A, Bot S, Karjalainen K, Bona C: Kinetics of generation and persistence of membrane class II molecules of a viral peptide expressed on foreign and self proteins. J Immunol 1996, in press.
60   Simitsek PD, Campbell DG, Lanzavecchia A, Fairweather N, Watts C: Modulation of antigen presentation by bound antibodies can boost or suppress class II MHC complex presentation of different T cell determinants. J Exp Med 1995;181:1957–1963.
61   Lanzavecchia A, Reid PA, Watts C: Irreversible association of peptides with class II MHC molecules in living cells. Nature 1992;357:249–252.
62   Unanue ER: Macrophages, antigen presenting cells and the phenomenon of antigen handling and presentation; in Paul WE (ed): Fundamental Immunology. New York, Raven Press, 1989, p 95.
63   Braciale TL, Braciale VL: Antigen presentation structural themes and functional variations. Immunol Today 1992;12:124–129.
64   Germain RN, Margulies DH: The biochemistry and cell biology of antigen processing and presentation. Annu Rev Immunol 1993;11:403.
65   Brumeanu T, Swiggard WJ, Steinman RM, Bona CA, Zaghouani H: Efficient loading of identical viral peptide onto class II molecule by antigenized immunoglobulin and influenza virus. J Exp Med 1993;178:1795–1799.

66   Brumeanu TD, Dehazya P, Bot S, Casares S, Wolf I, Bona CA: MHC class II presentation of a viral epitope assembled to carbohydrate moieties of self immunoglobulins does not require antigen processing, submitted.
67   Chain BM, Bou-Gharios G, Olsen I: Endopeptidase activities associated with the plasma membrane compartment of an antigen-presenting B cell. Clin Exp Immunol 1989;75:87–92.
68   Bently GA, Boulot G, Rittot MM, Poljak RJ: Three-dimensional structure of an idiotype-anti-idiotype complex. Nature 1990;348:254.
69   Fields B, Goldbaum FA, Ysern X, Poljak RJ, Mariuzza RA: Molecular basis of antigen mimicry by an anti-idiotype. Nature 1995;374:739–742.
70   Lanza P, Billeta R, Antonenko S, Zanetti M: Active immunity against the CD4 receptor by using an antiobdy antigenized with residues 41–55 of the first extracellular domain. Proc Natl Acad Sci USA 1993;90:11683–11687.

Constantin A. Bona, Mount Sinai School of Medicine, Department of Microbiology,
One Gustave Levy Place, Annenberg Bldg. 16–60, New York, NY 10029 (USA)

Subject Index

Antibody, *see also* Immunoglobulin A,
  Immunoglobulin G, Immunoglobulin M
  antigenization, *see* Arg-Gly-Asp (RGD)
    motif
  bound surface area 13, 14
  classes and nomenclature 88, 89
  computer modeling of structure 172,
    173
  concentration in plasma 88
  expression in Chinese hamster ovary
    cells 23, 24
  fusion with nonimmunoglobulins, *see*
    Fusion protein
  gene structure 129, 130
  hinge, *see* Hinge region
  monoclonal human antibodies, difficulty
    in production 19
  structure 57, 58, 88, 129, 159, 160
  therapeutic potential of engineered
    antibodies 130, 131
Antibody-dependent cell-mediated
  cytotoxicity, modulation through
  CD16 101, 102
Arg-Gly-Asp (RGD) motif
  adhesion protein association 159
  antigenization of antibodies
    adhesion and inhibition assay 165–169,
      171
    effect on peptide structure 172, 173
    principle 161
    procedure 163, 165
    selection of engineered sequence 161,
      163, 166, 172
    tumor therapy 175

Western blot analysis of antibody
  165, 166
expression on tumors 174, 175

Brain, targeting with IgG3-growth
  factor constructs 133, 137–139, 151,
  152

CD16 (FcγRIII)
  antibody-dependent cell-mediated
    cytotoxicity modulation 101,
    102
  IgG interactions 101, 102
  types 101
CD32 (FcγRII), IgG interactions
  100
CD64 (FcγRI), IgG interactions 99
Complementarity-determining region,
  epitope insertion
  difficulty 201, 202
  doubly chimeric molecules 186, 187
  effect on structure 186
  genetic engineering 184, 185
  internal-image immunoglobulin,
    definitive criteria 185, 186
  replacing vs inserting sequences 187

Disulfide bond, *see* Hinge region

Epitope
  immunogenicity on carrier molecules
    179, 180
  size 13
  synthetic peptide synthesis 179

Fab fragments
   binding affinity compared to whole
      antibody 25–27
   biological half-life 23
   construction for various HIV-1 epitopes
      C1/C2 epitope 37, 38
      CD4-binding site 29–31, 35
      CDR walking and affinity
         improvement 41–43
      gp41 39, 40
      heavy chain CDR3 homology 40, 41
      N-terminus 35, 37
      neutralization potency 43–45
      V1 loop 38, 39
      V2 loop 37
      V3 loop 38
Fab′γ module
   antibody constructs with Fcγ modules
      bispecific chimeric construct 69–71
      planning 68, 69
      preparation
         F(ab′)$_2$ 64
         Fab′ maleimide 66
         Fab′-SH 66
         Fab′(SH)$_5$ 64
         Fab′-SS-Py 64–66
   phage display 20–23, 29, 30
Fc receptor
   classes 95, 115
   FcγRI, see CD64
   FcγRII, see CD32
   FcγRIII, see CD16
   FcRn
      neonatal transport of IgG 103, 104
      role in IgG catabolism 104–106
   IgG interactions 94, 95, 99–102
Fcγ module
   antibody constructs with Fab′γ modules
      bispecific chimeric construct 69–71
      planning 68, 69
   glycosylation 113–115
   preparation
      Fcγ1 from IgG digestion 66, 67
      Fc-maleimide 68
      Fc-SH 68
      Fc-SS-Py 67, 68
Fusion protein
   antigenization of antibodies with
      RGD motif

adhesion and inhibition assay 165–169,
   171
effect on peptide structure 172, 173
principle 161
procedure 163, 165
selection of engineered sequence 161,
   163, 166, 172
tumor therapy 175
Western blot analysis of antibody 165,
   166
chimeric antibody construction with
   disulfide exchange
   bispecific chimeric construct 69–71
   planning 68, 69
complementarity-determining region,
   epitope insertion
   difficulty 201, 202
   doubly chimeric molecules 186, 187
   effect on structure 186
   genetic engineering 184, 185
   internal-image immunoglobulin,
      definitive criteria 185, 186
   replacing vs inserting sequences 187
cross-linking reagents 180, 182
enzymatically-glycosidic-mediated
   conjugation
   advantages 183, 184, 202
   efficiency 182
   principle 182
   specificity of attachment 182, 183
gene substitution in constant region
   131, 132
IgG3 constructs
   insulin-like growth factor/transferrin
      fusion
      blood-brain barrier permeability
         133, 137–139, 151, 152
      construction 133–135
      effector function 136, 137
      receptor binding 135, 136
   interleukin-2 fusion
      cell proliferation induction 141, 152
      complement-mediated lysis 142
      construction 141
      cytotoxicity 141
      enhancement of host antibody
         response 145, 146
      half-life 143, 152
      receptor affinity 141, 142, 152

tumor immunotherapy 139–141,
143–145, 152
polymers
complement activation 148, 149, 153,
154
construction 146, 148
Fc receptor binding 149
half-life 149–152
immunoadhesins 131
T cell antigen, chimeric
immunoglobulins
immunogenicity 190–193
immunopotency 193
kinetics of generation 193–195
molecular mechanisms of epitope
presentation 197, 199–201
persistence of peptide 194, 195
T cell recognition 188, 189
viral peptides 189–193

Glycosylation
engineered antibodies
cell lines 121–123
chimeric antibodies 122, 123
clinical requirements 121
costs of production 121, 123
site manipulation 124, 125
enzymatically-glycosidic-mediated
conjugation
advantages 183, 184
efficiency 182
principle 182
specificity of attachment 182, 183
Fc site 113
IgA sites 75,76
IgG
autoimmune disease patterns 119, 120,
124
carbohydrate structure 112
considerations in antibody
engineering 111, 112, 121–125
effector functions 102, 103, 111, 114, 115
effects
antibody-dependent cell-mediated
cytotoxicity 116–118
complement lysis 118, 119
Fc receptor binding
specificity 115–117
half-life of antibody 118

rheumatoid factor reactivity 119, 120
sites 112–114
structural consequences 114, 115
IgM 120, 121
gp41, *see* Human immunodeficiency virus-1
gp120, *see* Human immunodeficiency
virus-1

α-Helix, protein design 1
Hinge region
antibody constructs with disulfide
exchange
bispecific chimeric construct 69–71
planning 68, 69
disulfide bonds
alkylation of free sulfhydryl groups by
maleimides 62, 63
disulfide interchange 61, 62
functional types 59, 60
oxidation of free sulfhydryl groups 63
reduction 57
Fab'γ module preparation
F(ab')₂ 64
Fab' maleimide 66
Fab'-SH 66
Fab'(SH)₅ 64
Fab'-SS-Py 64–66
Fcγ module preparation
Fcγ1 from IgG digestion 66, 67
Fc-maleimide 68
Fc-SH 68
Fc-SS-Py 67, 68
genetic hinge 58
proteolytic cleavage 58, 59
structural hinge 58
structure 57, 58, 111
Human immunodeficiency virus-1
CD4 chimeric immunoglobulin
therapy 188
gp120
antibodies to CD4-binding site
Fabs 29–35, 37–39
neutralization of unpassaged
plasmavirus 32, 33
protection against infection in
mice 33–35
screening 28, 29
whole antibodies 31–33
epitope clusters 28

Human immunodeficiency virus-1,
gp120 (continued)
   functions in virus 27, 28
  monoclonal Fab fragments
    binding affinity compared to whole
     antibody 25–27
    construction for various epitopes
     C1/C2 epitope 37, 38
     CD4-binding site 29–31, 35
     CDR walking and affinity
      improvement 41–43
     gp41 39, 40
     heavy chain CDR3 homology 40, 41
     N-terminus 35, 37
     neutralization potency 43–45
     V1 loop 38, 39
     V2 loop 37
     V3 loop 38
    phage display 20–23
  neutralization assay 28, 30–32, 43–45, 48
  passive immunotherapy outlook 18, 19,
   24, 25
  vaccine assessment by phage
   display 45–47

Immunoglobulin A, *see also* Antibody
  allotypes 74, 75
  autoimmune disorders 74
  comparative biology 80, 81
  effector functions 79, 80
  expression system for molecular
   analysis 81, 82
  glycosylation 75, 76
  humoral protection in mucosal surface 73
  immunotherapy 82, 83
  J chain
   evolution 81
   function 76, 77
  polymerization 73, 76, 77
  secretory antibody
   receptors 73, 74, 78–80
   transport 73, 77, 78
  structure 74–76
Immunoglobulin G, *see also* Antibody
  aggregation 94
  biological half-life 104, 105
  catabolism and receptor binding 104–106
  Fc receptor interactions 94, 95
  flexibility in antigen binding 93, 94

functions 93–95
fusion constructs of IgG3
  insulin-like growth factor/transferrin
   fusion
    blood-brain barrier permeability
     133, 137–139, 151, 152
    construction 133–135
    effector function 136, 137
    receptor binding 135, 136
  interleukin-2 fusion
    cell proliferation induction 141, 152
    complement-mediated lysis 142
    construction 141
    cytotoxicity 141
    enhancement of host antibody
     response 145, 146
    half-life 143, 152
    receptor affinity 141, 142, 152
    tumor immunotherapy 139–141,
     143–145, 152
  polymers
    complement activation 148, 149, 153,
     154
    construction 146, 148
    Fc receptor binding 149
    half-life 149–152
glycosylation
  autoimmune disease patterns 119, 120
  carbohydrate structure 112
  considerations in antibody
   engineering 111, 112, 121–125
  effector functions 102, 103, 111, 114,
   115
  effects
    antibody-dependent cell-mediated
     cytotoxicity 116–118
    complement lysis 118, 119
    Fc receptor binding specificity
     115–117
    half-life of antibody 118
    rheumatoid factor reactivity 119, 120
  sites 112–114
  structural consequences 114, 115
isotypes 89, 93
neonatal transport 103, 104
recombinant antibodies and effector
  function 95–97
residues affecting binding
  CD16 101, 102

CD32 100
CD64 99
complement 97–99
sequence homology 89–91, 93
structure 89, 93, 111
Immunoglobulin M, *see also* Antibody
 glycosylation 120, 121
 polymerization 146, 150, 151
Immunotherapy, *see* Passive immunotherapy
Insulin-like growth factor, fusion constructs
 of IgG3
 blood-brain barrier permeability 133,
  137–139, 151, 152
 construction 133–135
 effector function 136, 137
 receptor binding 135, 136
Interleukin-2
 antitumor activity 139
 fusion constructs of IgG3
  cell proliferation induction 141, 152
  complement-mediated lysis 142
  construction 141
  cytotoxicity 141
  enhancement of host antibody
   response 145, 146
  half-life 143, 152
  receptor affinity 141, 142, 152
  tumor immunotherapy 139–141,
   143–145, 152
 half-life 139, 140, 143
Interleukin-6, minibody binding and
 screening 7–10
Internal-image immunoglobulin, *see*
 Fusion protein

Minibody
 bound surface area 13, 14
 circular dichroism spectroscopy 4, 5
 clinical applications 12, 13
 nuclear magnetic resonance 5
 optimization of binding 9, 10
 rationale for design 2
 rigidity of conformation 4
 screening
  interleukin-6 binding 7–10
  phage display 6, 7
  serine protease inhibition 10, 12
 solubility 4, 9
 structure 2, 4, 5

NS3 protease, minibody binding and
 screening 10, 12

Paratope, size 13
Passive immunotherapy
 efficacy against viruses 18, 19, 24, 25
 monoclonal human antibodies, difficulty
  in production 19
Phage display
 library size and binding affinity 6, 7
 monoclonal Fab fragments 20–23, 29, 30
 panning 20
 principle 6, 20
 vaccine assessment 45–47
pIgR
 gene 78
 IgA dimer binding 78, 79
 structure 78

Rational protein design, combination with
 screening 1, 2
Recombinant antibody, *see* Fusion protein
RGD motif, *see* Arg-Gly-Asp motif

β-Sheet, protein design 2
Sulfhydryl group, *see* Hinge region
Synthetic peptide, epitope synthesis 179

T cell antigen, chimeric immunoglobulins
 immunogenicity 190–193
 immunopotency 193
 kinetics of generation 193–195
 molecular mechanisms of epitope
  presentation 197, 199–201
 persistence of peptide 194, 195
 T cell recognition 188, 189
 viral peptides 189–193
Transferrin, fusion constructs of IgG3
 blood-brain barrier permeability 133,
  137–139, 151, 152
 construction 133–135
 effector function 136, 137
 receptor binding 135, 136
Tumor
 angiogenesis 174
 Arg-Gly-Asp motif-antigenized
  antibodies and immunotherapy 174,
   175
 immunotherapy goals 173, 174

Tumor (continued)
  interleukin-2 fusion with IgG
    cell proliferation induction 141, 152
    complement-mediated lysis 142
    construction 141
    cytotoxicity 141
    enhancement of host antibody
      response 145, 146
    half-life 143, 152
    receptor affinity 141, 142, 152
    tumor immunotherapy 139–141,
      143–145, 152
  metastasis 160, 166, 174

Variable domain, structure 2

Virus, *see also* Human immunodeficiency
  virus-1
  Fab binding affinity compared to whole
    antibody 25–27
  neutralization, in vitro vs in vivo 27
  passive immunotherapy outlook 18, 19,
    24, 25
  T cell antigen, chimeric immunoglobulins
    immunogenicity 190–193
    immunopotency 193
    kinetics of generation 193–195
    molecular mechanisms of epitope
      presentation 197, 199–201
    persistence of peptide 194, 195
    T cell recognition 188, 189